Exploratory Search

Beyond the Query–Response Paradigm

Synthesis Lectures on Information Concepts, Retrieval, and Services

Exploratory Search: Beyond the Query–Response Paradigm
Ryen W. White and Resa A. Roth
www.morganclaypool.com

ISBN: 9781598297836 paperback

ISBN: 9781598297843 ebook

DOI: 10.2200/S00174ED1V01Y200901ICR003

A Publication in the Morgan & Claypool Publishers series

SYNTHESIS LECTURES ON INFORMATION CONCEPTS, RETRIEVAL, AND SERVICES #3

Lecture #3

Series Editor: Gary Marchionini, University of North Carolina, Chapel Hill

Series ISSN Pending

Exploratory Search

Beyond the Query–Response Paradigm

Ryen W. White
Microsoft Research

Resa A. Roth

SYNTHESIS LECTURES ON INFORMATION CONCEPTS, RETRIEVAL, AND SERVICES #3

MORGAN & CLAYPOOL PUBLISHERS

ABSTRACT

As information becomes more ubiquitous and the demands that searchers have on search systems grow, there is a need to support search behaviors beyond simple lookup. Information seeking is the process or activity of attempting to obtain information in both human and technological contexts. *Exploratory search* describes an information-seeking problem context that is open-ended, persistent, and multifaceted, and information-seeking processes that are opportunistic, iterative, and multi-tactical. Exploratory searchers aim to solve complex problems and develop enhanced mental capacities. Exploratory search systems support this through symbiotic human–machine relationships that provide guidance in exploring unfamiliar information landscapes.

Exploratory search has gained prominence in recent years. There is an increased interest from the information retrieval, information science, and human–computer interaction communities in moving beyond the traditional turn-taking interaction model supported by major Web search engines, and toward support for human intelligence amplification and information use. In this lecture, we introduce exploratory search, relate it to relevant extant research, outline the features of exploratory search systems, discuss the evaluation of these systems, and suggest some future directions for supporting exploratory search. Exploratory search is a new frontier in the search domain and is becoming increasingly important in shaping our future world.

KEYWORDS

exploratory search, information seeking, augmenting human intellect

Preface

Search is a fundamental life activity. Mankind has become reliant on automated search technology to facilitate rapid access to pertinent information and to learn about the world. Exploratory search has emerged as an important research area with a focus on understanding and supporting searches that may result from ill-defined information needs, require explorative search strategies, or have personal development as a primary objective. The goal of exploratory search is to foster learning and investigation by capitalizing on innate human curiosities, moving beyond traditional information finding.

The failure to adequately differentiate exploratory search from other classes of information seeking has caused debate among academics and practitioners regarding its validity as a separate subdiscipline. This lecture takes a major step toward defining exploratory search and clarifying its relationship to information retrieval, information science, human–computer interaction, and psychology. The lecture aims to empower those interested in learning more about exploratory search, by providing them with knowledge about the current state of the field and future opportunities.

Contents

CHAPTER 1

Introduction

We shall not cease from exploration. And the end of all our exploring will be to arrive where we started and know the place for the first time.

T.S. Eliot
Little Gidding (1942, Part V: Lines 27–28)

1.1 OVERVIEW

Humans are explorers by nature, we seek to extend our knowledge by journeying beyond visible horizons. Through interaction with our environment, we aim to fulfill social and psychological needs to integrate with and learn about our world. As we explore, we gather information in order to develop complex intellectual skills such as comprehension, application, analysis, synthesis, and evaluation (Bloom, 1956) within a topical area, and through these skills facilitate self-actualization (Maslow, 1954).

Our desire to consume information exists in tension with how we should use it for our benefit. Significant advances in information technology in recent decades have afforded us the opportunity to use electronic information and compute cycles for human intelligence amplification (Ashby, 1956). Visionaries such as Vannevar Bush, J.C.R. Licklider, and Douglas Engelbart charted the course toward the use of information technology for the augmentation of human intellect. Bush (1945) envisioned the *memex* (or "memory extender"), an electromechanical device that an individual may use to read a large self-contained research library, and to add or follow associative trails of links and notes created by that individual or recorded by other researchers. This led to work on hypertext (Nelson, 1965), spatial hypertext (Marshall and Shipman, 1995) and ultimately the World Wide Web. Licklider (1960) pioneered real-time interactive computing and suggested that human–computer symbiosis would support decision making, the control of complex situations, or insight, by freeing the mind from mundane tasks. Engelbart (1962) proposed the enhancement of human intellect by increasing the capability of a human to approach a complex problem situation, gain comprehension to suit his particular needs, and to derive solutions to problems. He defined this increase in capability as "more-rapid comprehension, better comprehension, the possibility of gaining a useful degree of comprehension in a situation that previously was too complex, speedier solutions, better solutions, and the possibility of finding solutions to problems that before seemed insoluble." In fact, these are all reasonable metrics to evaluate the performance of such systems; evaluation will be discussed in Chapter 5.

Electronic corpora such as the World Wide Web, encyclopedias such as Microsoft Encarta,[1] and governmental databases are abundant sources of information. Automated search systems help users find information in such collections. The predominant retrieval paradigm these systems use is "query and response," where queries are issued by the user, and a set of potentially relevant items are offered in response. However, to develop complex intellectual skills, this lookup-based approach is insufficient; it yields only candidate starting points for learning, not the complete set of items required for significant cognitive development. People are forced to consume information independently from search systems, navigating based on their information needs and items' information scent.

Bush, Licklider, and Engelbart all advocated for a symbiotic relationship between humans and machines that would involve changes in the way humans tackle complex problems; machines could be viewed as cognitive prosthestics. At the time of conception, four to five decades ago, computing technology was insufficiently advanced to implement many of these revolutionary ideas. Computer manfacturers could not mass produce systems at sufficiently low price to empower the general population. However, much has changed in the manufacture and, more importantly, costs of high-performance computing technology. We are now in a strong position to create the enlightened society that Bush, Licklider, Engelbart, Nelson, and others envisioned.

The goal of increasing intelligence through the development of cognitive prosthestics is admirable. However, a potential obstacle to the success of these agents is the rapid growth of new information. Information overload has become a significant problem for many of us,[2] especially given our seemingly insatiable thirst for knowledge:

> What information consumes is rather obvious: it consumes the attention of its recipients. Hence a wealth of information creates a poverty of attention, and a need to allocate that attention efficiently among the overabundance of information sources that might consume it (Simon, 1971, pp. 40–41).

[1] http://www.microsoft.com/encarta

[2] For example, in 2006, the amount of digital information created and replicated worldwide was estimated to be 161 exabytes (an exabyte is a billion gigabytes, or 10^8 bytes), with 2008 estimates exceeding 500 exabytes.

To ensure that only the most pertinent information reaches us at any given time, augmenting agents must be employed to control our rate of information consumption. Although we are *informavores* (Miller, 1983; i.e., a species that hungers for information in order to gather it to adapt to our world; Pirolli, 2008), only a small amount of the information we encounter is actually relevant to our current activity. To reduce the challenges posed by information overload, information filtering tools such as recommender systems, use a personalized (social/collaborative) profile to remove redundant or unwanted information from information streams (Loeb and Terry, 1992). To obtain information that is relevant to their current activity, people usually seek out information from repositories such as libraries, newspapers, and the World Wide Web. Information seeking is a fundamental human activity comprising systematic and opportunistic elements that provides many of the "raw materials" for planned behavior, decision making, and the production of new information products (Marchionini, 1995).[3] Wireless computing technology facilitates anytime, anywhere information access. The snowballing effect of pervasive access to information, coupled with the expected growth in the range of search task types being attempted, brings new challenges to information-seeking theory and design of information-seeking support systems (Marchionini and White, 2009). One emerging class of information seeking—known as exploratory search—requires search systems to help users clarify vague information needs, learn from exposure to information in document collections, and investigate solutions to information problems. Systems supporting exploratory search facilitate intellectual growth and long-term personal/professional development, as well as task completion and user satisfaction.

1.2 BACKGROUND

For decades, researchers in the information retrieval (IR) research community have studied how people manipulate, store, retrieve, and disseminate information in settings such as libraries, commerical organizations, and the personal computer. They have developed automated search tools to help people locate relevant information in an efficient manner (e.g., Van Rijsbergen, 1979; Salton and McGill, 1983). The lookup-based retrieval model that has been used in the IR community to represent search activity can be characterized as shown in Figure 1.1.

The components of this model are (1) the collection being searched, (2) a representation of the documents that stored the collection (usually as an inverted index for rapid document lookup),

[3] We regard information seeking as subtly different from information retrieval (IR). In IR, the target is typically known, its existence confirmed prior to query issuance, and the user's task is to create the well-formed query that will retrieve it. In IR, the task is to retrieve relevant documents at the top of the ranked list. In contrast, in information seeking, there is uncertainty over whether the information being sought exists and whether the searcher, working in synergy with the system, will be able find it.

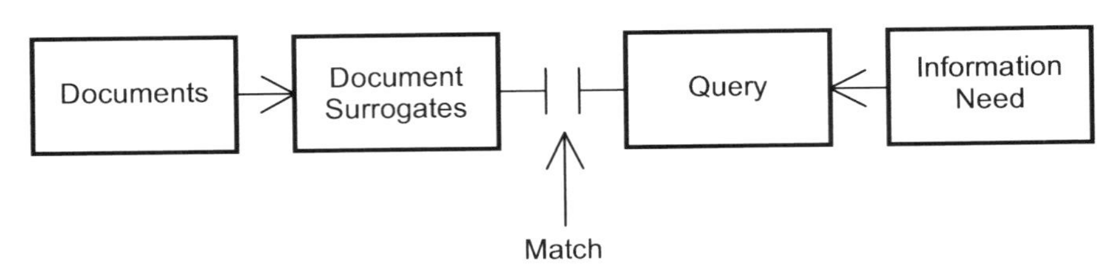

FIGURE 1.1: Lookup-based IR model (based on Bates, 1989).

(3) the underlying information need of the searcher, and (4) a query statement (in textual form provided by the searcher at query time). This is the dominant interaction model currently used in the development for database management systems and in major commercial Web search engines offered by companies such as Google,[4] Yahoo!,[5] and Microsoft.[6] To use these engines, searchers generally provide a textual query via the homepage of the search engine or the Web browser, and a ranked list of captions comprising titles, snippets, uniform resource locators (URLs), and other relevant information (e.g., Web page size and most recent page modification timestamp) are returned for inspection and subsequent document selection. Lookup tasks are usually suited to analytical search strategies that begin with carefully specified queries and yield precise results with minimal need for result set examination and item comparison (Marchionini, 2006a). As such, systems supporting such tasks are best suited for fact-finding or question-answering scenarios.

The lookup-based model has promoted our understanding of IR in many ways (e.g., the basis under which systems are evaluated at the Text Retrieval Conference; Harman, 1993; Voorhees and Harman, 2005). Under this paradigm, the query is treated as a one-time conception of the searcher's information need. Although the assumptions regarding user interaction are useful abstractions to simplify IR system research, real-life searches typically contain multiple query iterations, postquery browsing, and detailed result examination, all of which are not captured in this model. Indeed, as Kuhn (1970) noted, major models that are central to a field eventually begin to show inadequacies as testing leads to improved understanding of the processes being studied. This is increasingly true of the lookup-based model as a basis for information-seeking research, where humans and their search context have emerged as important participants in the search process (Bates, 1989; Ingwersen and Järvelin, 2005).

In recent years, the traditional lookup-based interaction model has been attacked (e.g., Bates, 1986a,b; Belkin et al., 1982a,b; Ellis, 1984; Ingwersen, 1992; Kuhlthau, 1993; Marchionini,

[4] http://www.google.com

[5] http://www.yahoo.com

[6] http://www.live.com

1995). This negative sentiment stems from its inability to fully represent how humans interact with search systems, the potential dynamism of information needs during a search session, and its ignorance of important factors such as task context and information use. Most of the proposed information-seeking models characterize information seeking as a process that occurs over time across many search episodes using disparate resources. Information seeking is typically intertwined with many other activities, and it is common for users to be engaged in multiple information-seeking tasks simultaneously.

Interactive information retrieval (IIR) focuses on how people use IR systems to retrieve information (i.e., sought information exists, and IIR studies information finding; Ruthven, 2008). Human–computer information retrieval (HCIR; Marchionini, 2006b) is emerging as an important subdiscipline focused on the role of the searcher and their context on the search process. Information seeking depends on the cognitive representation (mental model) of a system's features, which is largely determined by the conceptual model that designers provide through the human–computer interface. Other determinants of successful retrieval include the user's knowledge of the task domain, information-seeking experience, and physical setting (Marchionini and Shneiderman, 1988). To support more effective search interactions, HCIR leverages advancements in user interface technology and an improved understanding of users' search strategies developed by the information scientists studying library patrons and their interactions with reference librarians.

Interactive search systems offer support such as relevance feedback (RF), information visualizations, and query suggestions. RF (cf. Salton and Buckley, 1990) allows searchers to provide implicit or explicit feedback about relevant information and uses these judgments to enhance subsequent searches. Information visualizations (e.g., Card et al., 1999) use graphical techniques to visually represent large-scale collections of non-numerical information and help searchers attain new insights in support of decision making or other related complex mental activities. Query suggestions (e.g., Koenemann and Belkin, 1996) provide recommendations about which query terms to add or queries to issue that assist with the challenging process of query formulation.

Oddy (1977) and Belkin and colleagues (1982a,b), among others, questioned the requirement for searchers to represent their information needs in a query understandable by the system. Indeed, systems such as I^3R (Croft and Thompson, 1987), Bead (Chalmers and Chitson, 1992), and Ostensive Browser (Campbell and Van Rijsbergen, 1996) offer "query-less" interfaces, where searcher needs are conveyed by means of examples from their browse behavior rather than textual descriptions. Research on implicit feedback (Joachims et al., 2005; Kelly and Belkin, 2004; Kelly and Teevan, 2003; White et al., 2005b) has shown that interaction behavior (mainly document retention activities such as saving, bookmarking, and printing, as well as search engine result page click-through) can be used to build enhanced representations of information needs for use in query refinement or future retrieval.

It can be argued that retrieval alone is generally sufficient when the need is well-defined in the searcher's mind. However, when information is sought to address broad curiosities, for learning, decision making, and other complex mental activities that take place over time, retrieval is necessary but not sufficient. In this lecture, we focus on exploratory search, a concept that covers such information-seeking requirements. We adopt the following definition of exploratory search:

> Exploratory search can be used to describe an information-seeking problem context that is open-ended, persistent, and multi-faceted; and to describe information-seeking processes that are opportunistic, iterative, and multi-tactical. In the first sense, exploratory search is commonly used in scientific discovery, learning, and decision-making contexts. In the second sense, exploratory tactics are used in all manner of information seeking and reflect seeker preferences and experience as much as the goal (Marchionini, 2006a).

In exploratory search, people usually submit a tentative query to navigate proximal to relevant documents in the collection, then explore the environment to better understand how to exploit it, selectively seeking and passively obtaining cues about their next steps (White et al., 2006a). Information exploration is a broad class of activities where new information is sought in a defined conceptual area. Exploratory search can be considered a specialization of information exploration; exploratory data analysis (Tukey, 1977) is another such specialization. As illustrated with the previous quotation, exploratory search is also a specialized form of information seeking, in terms of problem context and/or search strategies employed. In many ways, exploratory search is as much about the journey through the information space as the destination (i.e., the relevant document with the sought answer). The answer may not be immediately obvious. In exploratory searches, it may only emerge after analysis of the information gathered during one's journey (sometimes spanning multiple days, weeks, or even months). Exploratory search can have a profound impact on users' personal development, as it surpasses knowledge acquisition en route toward higher-level learning objectives.

People conducting exploratory searches—referred to hereafter as exploratory searchers—need systems to support their specific search activities. Exploratory search systems (ESSs) capitalize on new technological capabilities and interface paradigms that facilitate an increased level of interaction with search systems. ESSs help people dynamically manage, analyze, and share sets of retrieved information. The information needs of people grappling with chronic illness, work teams creating complex solutions or products, learners studying complex material over time, families making long-term plans, scientists investigating complex phenomena, and hobbyists tracking developments over a life time are well-served at only the most superficial levels by existing Web search engines (Marchionini and White, 2009). ESSs address this shortcoming by providing search solutions that empower users to go beyond single-session lookup tasks. It is during complex search scenarios that information seekers require support from systems that extends beyond the provision of search results.

Examples of ESSs include information visualization systems, document clustering and browsing systems, and intelligent content summarization systems. ESSs go beyond returning a single document or answer in response to a query and instead aim to instigate significant cognitive change through learning and improved understanding. ESSs support aspects of sense-making, information foraging, and berrypicking. Sense-making (Dervin, 1977, 1998): through information visualization and other depictions, ESSs help create situational awareness and understanding in support of decision making. Information foraging (Pirolli and Card, 1995, 1999): ESSs support the exploration and identification of information patches and maximal information gain. Berrypicking (Bates, 1989): through query refinement support and dynamic queries, ESSs support the query evolution over time, and through RF and scratchpads, ESSs help people gather information in chunks rather than in a single result set. For example, browsing is a serendipitous activity that can be attractive to users who may benefit from the extraneous information (Marchionini and Shneiderman, 1988). ESSs help users engaged in browsing maximize their rate of information gain, make decisions about which navigational paths to follow, and understand the information they encounter. In addition, through interface features such as dynamic queries (Ahlberg et al., 1992), ESSs can help users see the immediate impact of their decisions on visualizations of the data such as starfield displays, cartograms, and histograms.

1.3 EMERGING INTEREST IN EXPLORATORY SEARCH

In recent years, there has been a growing interest in exploratory search. The interdisciplinary Exploratory Search Interfaces (XSI) workshop held at the University of Maryland at College Park in June 2005, organized by Ryen White, Bill Kules, and Ben Bederson initiated research in this area (White et al., 2005c). It united researchers from the human–computer interaction, psychology, IR, and information science communities for a discussion of the issues related to the design of exploratory search interfaces. The overarching aim of the event was to create a working definition of exploratory search. This was appropriate and necessary given the infancy of the subdiscipline at that point in time. The XSI workshop spawned special issues of the *Communications of the Association for Computing Machinery (ACM)* in April 2006 (targeting tools to support exploratory search interaction) and the *International Journal of Information Processing and Management* in March 2008 (targeting the evaluation of exploratory search systems). It also led to workshops at the 2006

ACM SIGIR Conference on Research and Development in Information Retrieval (revisting ESS evaluation) and the 2007 *ACM SIGCHI Conference on Human Factors in Computing Systems* (revisiting ESS interface design). In 2008, the United States National Science Foundation (NSF) sponsored an event at the University of North Carolina in Chapel Hill that brought together leaders from academia and industry to devise a research agenda for the future of information-seeking support systems, with a large emphasis on exploratory search activities and complex search scenarios. Three mutually interdependent requirements emerged from the workshop: (1) more robust models of human–information interaction; (2) new tools and services to meet the expanding expectations and more comprehensive information problem space; and (3) better techniques and methods to evaluate information seeking across platforms, sources, and time.

The workshops and special issues have created opportunities to build a diverse community of interest in this area. They have also been vital in identifying many of the important issues, creating working definitions for exploratory search, helping practitioners develop ESSs to support exploratory searchers, and devising metrics and methodologies to evaluate ESSs. In addition to elaborating on the above ideas, this lecture will demonstrate the importance of exploratory search, discuss how exploratory searches are being supported by systems currently, including what is lacking. Also, we predict where searchers can expect to see major developments in search technology in the future, as exploratory search systems become more prevalent.

1.4 STRUCTURE OF THE LECTURE

The remainder of this lecture is structured as follows: Chapter 2 offers a definition of exploratory search that covers the problem context and the search process. Chapter 3 presents related work drawn from the IR, information science, psychology, and human factor literature. Chapter 4 presents a set of features that users should expect from exploratory search systems and relates them to existing search systems. Chapter 5 discusses issues around the evaluation of ESSs, in particular what researchers should consider when planning ESS evaluation, and Chapter 6 projects future directions for exploratory search.

* * * *

CHAPTER 2
Defining Exploratory Search

The definition of exploratory search is complex and multifaceted. Almost all searches are in some way exploratory. Although there may be circumstances where exploratory strategies are used continually to allow people to discover new associations, kinds of knowledge, and decision making, they are often motivated by a complex information problem, a poor understanding of terminology and information space structure (White et al., 2006a), and a desire to learn. In this chapter, we propose a definition of exploratory search. We first focus on two important elements: the problem context and the search process and then combine them in a model of exploratory search.

To assist with exposition, this chapter begins with an example of an exploratory search:

> Meet George, a U.S. citizen planning a vacation to the south of France. He has never been to Europe and wishes to experience French culture as an important aspect of his journey. To this end, he wants to rent a villa in a remote village. First, George uses a Web search engine to find out whether this is possible. He encounters a website that offers villa rental in Provence. After investigating Provence and deciding that he likes the region, he looks up villa rental prices and decides that he needs to adjust his goals. The only available villa rentals during his desired travel window are prohibitively expensive, so George decides to book a hotel in Marseille instead. He searches for accommodation with a minimum rating of three stars, studies the websites of a few hotels, decides on a hotel that meets his needs, and proceeds to make a reservation. Following the booking, he needs to investigate transportation options, learn more about French customs and cuisine, and identify sightseeing destinations. He has much to learn and investigate before his trip even begins.

As illustrated in this example, exploration is an important aspect of many search processes. However, it is not only the act of exploring that makes a search exploratory; it also must include complex cognitive activities associated with knowledge acquisition and the development of intellectual skills. Learning is an important mental function reliant on the acquisition of knowledge and supported by perceived information. It leads to the development of new capacities, skills, values, understanding, and preferences. Once a person has acquired information and internalized it, such that they understand its meaning, translation, interpolation, and interpretation, they may then apply that knowledge in new domains and pursue higher-order learning activities such as analysis, synthesis, and evaluation (Anderson and Krathwohl, 2001; Bloom, 1956). Exploratory searchers utilize a combination of searching and browsing behavior to navigate through (and to) information that helps them develop powerful cognitive capabilities and leverage their newly acquired skills to address open-ended, persistent, and multifaceted problems. Searching to learn includes decision making, and professional and life-long learning. It also includes social search to find communities of interest (e.g., via social network systems; Marchionini, 2006a).

As suggested earlier, exploratory search can describe either the problem context that motivates the search or the process by which the search is conducted (Marchionini, 2006a). These two elements are tightly coupled; the resolution of vague or complex information problems requires exploratory search behaviors. Exploratory search covers a broader class of search activities than traditional IR and IIR, which targets query-document matching under the assumption that relevant information exists and that a well-formed query statement will retrieve it from the collection. Information visualization focuses on the visual representation of large collections to help people understand and analyze data. Information visualization is an important tool to support exploratory searches; however, it does not target information seeking or information use.

People engaged in exploratory searches are generally: (1) unfamiliar with the domain of their goal (i.e., need to learn about the topic in order to understand how to achieve their goal); (2) unsure about the ways to achieve their goals (either the technology or the process); and/or even (3) unsure about their goals. Exploratory search is a specialization of information seeking, which describes the activity of attempting to obtain information through a combination of querying and collection browsing. Affective and cognitive uncertainties are persistent characteristics in information seeking and, in particular, exploratory search. Indeed, Wilson (1999) refers to uncertainty during information seeking as an ever-present, unpleasant factor. Uncertainty is a natural user experience within the process of information seeking and acquiring meaning. It can give rise to feelings of doubt, confusion, frustration, and anxiety (Kuhlthau, 2004). Kuhlthau's model of the information search process portrays information seeking as a process of construction, with uncertainty decreasing as understanding increases (1991, 2004).

Increased uncertainty indicates a zone of intervention for human intermediaries such as reference librarians and system designers. Growing uncertainty is also an important part of exploratory search. The creativity, innovation, and knowledge discovery that is often necessary as part of exploratory searches requires traveling beyond what is known by the user. In a similar way to research practice, exploratory search involves original thought, lateral thinking, and serendipity (Bawden, 1986; Foster and Ford, 2003). The complexity of research practice leads to a nonlinear, dynamic process involving a tacking back and forth between deduction and induction (Budd, 2004). It involves balancing divergent thinking with the convergence of ideas (Ford, 1999). The processes of exploring and working with information are critical for building connections, discovery, and creativity. These processes rely on the effective provision, processing, and manipulation of information at all stages of an exploratory search.

2.1 PROBLEM CONTEXT

Searches are often motivated by an incompleteness (Ingwersen, 1992; Mackay, 1960; Taylor, 1968) or a "problematic situation" (Belkin, 1982a,b) in the mind of the searcher that develops into a desire for information. When a search begins, a searcher's state of knowledge is in an "anomalous state," and they have a gap between what they know and want to know. The gap is a situation-driven phenomenon, known as their information need. Exploratory searches may also be driven by curiosity or a desire for personal development; a user may only wish to learn more about a particular subject area to increase their knowledge rather than solve an information problem.

Exploratory searches often involve complex situations. Engelbart (1962) suggested that these situations include "the professional problems of diplomats, executives, social scientists, life scientists, physical scientists, attorneys, designers—whether the problem situation exists for 20 minutes or 20 years." He advocated for human–machine symbiosis during the resolution of complex situations and emphasized that this should not involve "isolated clever tricks that help in particular situations," but instead, "a way of life in an integrated domain where hunches, cut-and-try, intangibles, and the human 'feel for a situation' usefully coexist with powerful concepts, streamlined terminology and notation, sophisticated methods, and high-powered electronic aids."

The problem context in exploratory search is ill-structured, and users require additional information from external sources to clarify their goals and actions (Simon, 1973). Exploratory searchers are engaged in weak problem solving (Newell and Simon, 1972) with a lack of prior domain knowledge and/or unclear or unsystematic steps through the information space.[1] In information seeking, complex situations or tasks are often framed as wider information tasks involving problem solving (Attfield et al., 2003; Byström and Järvelin, 1995; Kuhlthau, 1993; Vakkari, 1999; Wilson, 1999). Ingwersen and Järvelin (2005) defined models of the tasks at varying levels of abstraction. The work task, viewed as the catalyst behind search activity, provides a problem context within which the searcher operates. Within the context of a single work task, users generally perform a number of smaller search tasks, designed to reach their goal incrementally. As part of this process, users must

[1] This contrasts with strong problem solving, where people research problems that are well-defined, systematic, and routine (Newell and Simon, 1972).

divide the larger work tasks into smaller tasks and tackle each in sequence or, if possible, in parallel. However, for work tasks that are complex or poorly defined, it can be difficult for users to divide the task into manageable chunks, since the information required to accomplish that task cannot be determined in advance (Byström and Järvelin, 1995; Vakkari 1999). These are areas where exploratory search systems can help users develop an improved knowledge of the task environment and, hence, facilitate more effective search task selection.

During exploratory searches, it is likely that the problem context will become better understood by the searcher, allowing them to make more informed decisions about interaction or information use. The recognition and acceptance of an information problem typically resides at the beginning of the information-seeking process (e.g., Ellis, 1989; Marchionini, 1995; Wilson, 1997). The problem can be internally motivated (e.g., curiosity) or externally motivated (e.g., an assignment). It may be characterized by a gap (Dervin, 1977), a visceral need (Taylor, 1968), an anomaly in a searcher's knowledge state (Belkin, 1982a,b), as a defect in a mental model, or as an unstable collection of noumena (Marchionini, 1995). Once the problem has been accepted, it must then be understood and defined. To do so, it must be limited, labeled, and a framework for the answer constructed. Taylor (1968) referred to this as the "conscious need." During this process, attributes of candidate solutions emerge that will ultimately guide user interaction behavior. This process leads to the development of Taylor's "formalized need" and the possible articulation of an information-seeking task. The user defines the problem internally as a task with properties that allow progress to be judged and a search strategy to be selected. The problem definition phases are an important part of exploratory search (perhaps even more so than in other problem contexts). The answer framework may still be poorly defined or highly variable in exploratory searches, but it is expected that a structure exists upon which an answer can be constructed.

The problem solution can be constructed from information within relevant documents and knowledge accumulated during the search, including the examination of partially relevant and irrelevant documents. The information need derived from the problem is prone to develop during the search and evolve from an initial, vague state into one known and understood by the searcher (Ingwersen, 1994). As the information need evolves, the searcher's ability to articulate query statements and identify relevant information increases based on their improved level of problem comprehension (Belkin, 2000).

Evidence from a number of studies on information-seeking behavior (Harter, 1992; Spink et al., 1998; Tang and Solomon, 1998) has shown that information needs are transient and developing. In exploratory searches, the problem context may remain undefined or in significant flux for much of the search session. There may also be periods of heightened uncertainty and confusion as people discover new information and assimilate knowledge. Tools to support exploratory search should help users define the problem, make sense of encountered information throughout the current session and across multiple sessions, and handle uncertainty and confusion by providing progress updates, explanations for system actions, and summaries of major themes present in encountered information.

Marchionini (2006a) suggested that key components of the exploratory search process are learning and investigation. To search in advancement of one's knowledge has been established as an important motivator behind information-seeking activities (e.g., Belkin, 1982a,b; Mackay, 1960; Taylor, 1968). The learning associated with exploratory search systems is subtly different. Rather than searching to close a gap in one's knowledge (where the gap may be known or its presence at least identified to the user at the outset of the search), the goal in exploratory searches may be less clearly defined; learning in exploratory search is not only about knowledge acquisition, but rather the development of higher-level intellectual capabilities within a particular subject area (e.g., application, synthesis, evaluation). The purpose of exploratory search is typically to create a knowledge product (e.g., a research paper) or shape an action (e.g., choosing a medical treatment; Pirolli, 2009).

2.2 SEARCH PROCESS

Learning searches involve multiple query iterations and return sets of items that require cognitive processing and interpretation. Much of the search time in learning tasks is devoted to examining and comparing results, as well as reformulating queries to discover the boundaries of key concept definitions. Learning search tasks are best suited to combinations of browsing and analytical strategies, with lookup searches embedded to locate the correct neighborhood for browsing.

Marchionini (2006a) proposed a set of search activities associated with an exploratory search process and separated the activities related to exploration from lookup searches, handled by traditional search technologies such as Web search engines. Figure 2.1 illustrates the search activities associated with lookup and exploratory searches.

Although the activities are shown separately in Figure 2.1, there is generally interplay between them (e.g., lookup searches are embedded in learning or investigation, learning is an important part of investigation). Lookup searches generally involve the retrieval of single answers (e.g., a single piece of information satisfies a known item search, fact retrieval, or question answering; a single Web page satisfies a navigational query submitted to a Web search engine), as seen in Figure 1.1. The majority of today's search systems handle lookup searches well, given the significant investment in ranking technologies and instant answers (e.g., weather forecasts or stock quotes) by major search companies such as Google, Yahoo!, and Microsoft. However, activities associated with exploratory searches require more involvement from the user, more synergy between user and search system, and more functionality from the system, extending beyond just query specification and result presentation. To

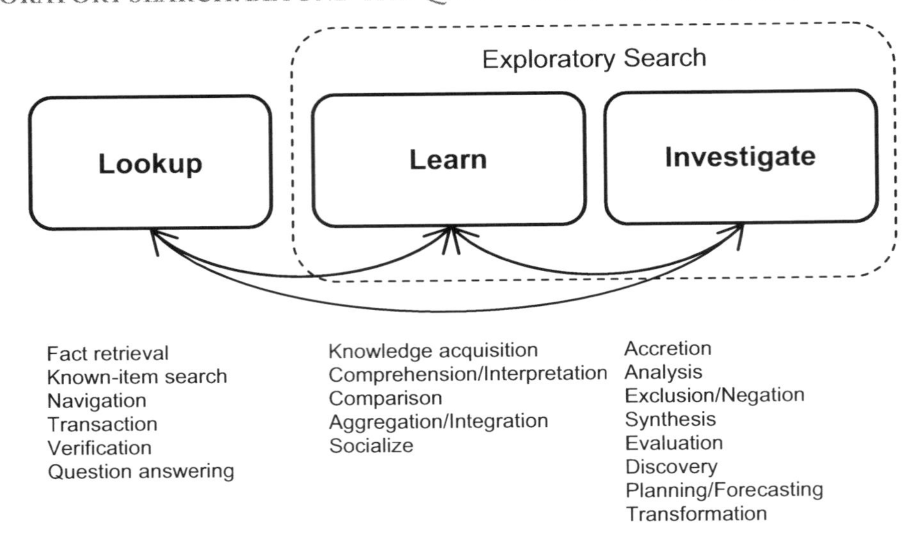

FIGURE 2.1: Exploratory search activities (based on Marchionini, 2006a). Arrows signify interaction between activities.

illustrate exploratory search, Marchionini (2006a) uses the act of social searching, where people try to find communities of interest or discover new friends on social network systems such as Friendster.[2] The search time in such tasks is spent examining and comparing results, and issuing and reformulating queries. Although social search captures some of the behaviors associated with exploratory searches, the learning objective is not ambitious. Systems tailored to supporting exploratory search processes should help instigate significant cognitive change and user development. This can only result from an extended learning process spanning multiple queries or search sessions rather than a single result or set of results offered by a system in response to a user's query.

Marchionini's model describes exploratory search at an intellectual level, derived from many of the educational objectives of Bloom's taxonomy (1956). However, the model does not examine the interaction behaviors that are likely associated with exploratory search activities. For example, exploratory searchers may exhibit a behavior akin to "wayfinding" (a concept borrowed from urban planning; Lynch, 1960), where they naïvely traverse the information landscape with no prior knowledge of the whereabouts of the information target, if a target exists. Wayfinding tasks generally require the navigator be able to conceptualize the space as a whole. This is analogous to what Thorndyke and Goldin (1983) refer to as survey knowledge. For example, a scientist visualizing data sets computed off-line may have no preconception of the shape or organization of the data. Therefore, wayfinding

[2] http://www.friendster.com

assistance requires support for both exhaustive and directed searches and must facilitate topological knowledge acquisition (i.e., help users learn about the location of information objects and paths through the information space). Exploratory searchers navigating an unfamiliar document collection may need similar assistance. Wayfinding is an area where trails followed by previous "trailblazing" users can help the current user (Bush, 1945; Wexelblat and Maes, 1999; White et al., 2007).

Serendipitous browsing stimulates analogical thinking, and users can relate their experiences to other comparable situations. Exploratory searches may be more concerned with recall (maximizing the number of possibly relevant objects that are retrieved) than precision (minimizing the number of possibly irrelevant objects that are retrieved). Thus, they are not well supported by today's Web search engines that are highly tuned toward precision in the first page of results. The principle of least effort (Zipf, 1949), applied in the information-seeking context, suggests that a searcher will tend to use the most convenient search method, in the least exacting mode available, and will stop searching when minimally acceptable results are found (Mann, 1987). Although this is often regarded as a guiding principle in information-seeking research, it is less applicable for exploratory searches. As stated earlier, exploratory searches are as much about the journey (and the learning that occurs) as the destination, if a destination exists. Systems that accelerate learning and promote topic coverage will help users assimilate knowledge more efficiently, but it is unlikely that users will simply terminate an exploratory search once relevant information fragments have been encountered. For example, if multiple sources of evidence are required, it is likely that users will need to validate these sources to determine their reliability before concluding.

Distinctions among different types of search activities suggest that lookup searches lend themselves to formal turn-taking, where the searcher poses a query, and the system performs the retrieval and returns results. The human and system take turns in retrieving the best result. However, exploratory search requires human participation in a continuous and exploratory process. This may involve the application of dynamic query filters to adjust the result presentation in real time (Ahlberg et al., 1992), dramatic evolution of information needs over the course of the search, and fundamental shifts in understanding.

Information seeking as berrypicking (Bates, 1989) is an influential metaphor and conceptual framework when considering information need evolution. Users often start with some vague information need and iteratively seek and select fragments of information that cause the information need and behavior to evolve over time; there is no one path of behavior to a single best query and retrieval set. Bates observed that during berrypicking, library users employed a wide variety of information navigation strategies, such as footnote chasing, citation chaining, reviewing a journal series, browsing entire areas at different levels of generality, and browsing and summarizing works by author. These existing information-seeking strategies need to be supported by system features and user interface designs, bringing humans more actively into the search process.

2.3 MODELING EXPLORATORY SEARCH BEHAVIOR

In this chapter, we have described two important elements of exploratory search: the problem context and the search process. The problem context is an important motivating factor, but is also highly dynamic in exploratory search scenarios. Over the course of an exploratory search, this dynamism may decrease as topic familiarity grows and user knowledge increases. This makes subtask identification more straightforward and the identification of pertinent information easier. Supporting the gathering and re-representation of information—as is common practice in sense-making (Dervin, 1977)—helps reduce the uncertainty inherent in the problem context. Search strategies that are exploratory in nature (e.g., berrypicking, information foraging) can be used for this task, but this need not always be the case. It is possible for a user to better define the problem context through systematic learning mechanisms such as hypothesis formulation and testing, as in exploratory data analysis (EDA; Tukey, 1977). In many respects, exploratory search is similar to EDA, especially during the early stages where the interaction between the perceived problem context and the information encountered occurs most rapidly. In EDA, the role of the researcher is to explore the data in as many ways as possible until a plausible "story" of the data emerges. In some respects, the researcher is a detective, collecting evidence and clues related to the central question of the case. This is also true of exploratory searchers, who are motivated to search by the problem context, although the relevance of encountered information to this context may not be immediately apparent. Relevance depends on the stage in the search and the searcher's level of domain knowledge, among other factors.

There are two main activities that reside in an exploratory search episode: exploratory browsing and focused searching. Exploratory browsing exposes users to collection content to help relate the problem context to similar documented experiences and promote information discovery. Focused searching may include some degree of navigation, but is generally intended to help the user follow a known or expected trail rather than forging new ground. Effective exploratory search systems will maintain a balance between analytical and browsing activities and support a symbiotic search relationship between searcher and system.

2.3.1 Exploratory Browsing

Browsing is defined as movement in a connected space (Kwasnik, 1992).[3] In order to browse effectively, people undertake certain actions beyond basic scanning (e.g., timely omission of irrelevant

[3] Kwasnik (1992) also identifies six activities that play a role in browsing: (1) persistently orienting to the environment; (2) marking of potentially relevant items for potential second consideration; (3) identification or recognition of potentially relevant or definitely not relevant items; (4) resolution of anomalies; (5) comparison between items that serve to orient, identify, and solidify purposes and aims; and (6) transitions from one item to another.

information) and respond to interesting phenomena. Exploratory browsing within document collections is performed by exploratory searchers to better define their information needs and to promote new ideas and cognition based on observed content. On the Web, this browsing activity occurs between hyperlinked pages and also among images in a digital library or through passageways in a virtual world. Browsing may be a hypothesis-generation activity, whereby hypotheses are generated about the causes of observed phenomena or the best ways to resolve an information problem. During hypothesis generation, users will visit multiple documents to better understand what information is available and familiarize themselves with the topic.

Concepts of browsing in information-seeking research have become increasingly sophisticated (e.g., Ingwersen and Wormell, 1989; Noerr and Noerr, 1985; Wade and Willett, 1988). As Ellis (1989) notes, browsing is an important part of standard information searching; he calls it "semidirected or semistructured searching" when used this way. He recommends that browsing of a variety of types of information, e.g., contents pages, lists of cited works, subject terms, should be made available in automated systems. Ellis further argues that since the user is conducting the browsing, we therefore do not have to design a cognitive model of user browsing into the system, and providing browsing features should be relatively simple. Bates (2004) suggests that browsing consists of a series of glimpses, leading to further or closer exploration of the documents viewed. In this way, browsing consists of numerous stops and starts, involving some reading or surveying, alternated with other actions, such as sampling and selecting.

Along with browsing, the collection to find relevant documents, to improve topic knowledge, or to make serendipitous discoveries, another key activity in exploratory search is focused searching. Although this may not comprise as much of exploratory search behavior as browsing, it is an important element in helping searchers resolve their problems.

2.3.2 Focused Searching

In focused search, people query the document collection, examine search results and documents in close proximity to search results, and extract relevant information to meet their goals. Searchers engaged in focused search may require analytical support for query specification and refinement, and for the selection of search results and postquery navigation paths. During focused searching, the user may have a clear sense of their information goals and the trails to follow to attain them. Searchers also may test the hypotheses generated during the hypotheses-generation activity of exploratory browsing.

The need for users to exhibit more than a few postquery interactions is linked to the inability of systems to fully understand the information needs of their users. As previously suggested by Teevan and colleagues (2004), the "perfect" search engine (i.e., a search engine that returns exactly what is sought, given a well-specified information need) may address some problems, but there may be circumstances where users are unable to specify their information needs at a level to make systems effective. Instead, it has been observed that users exhibit a style of interaction known as orienteering (O'Day and Jeffries, 1993). With orienteering, the search engine is used to transport users to a part of the information space containing potentially relevant documents. Users then rely on their recall and recognition skills to locate relevant information. Postquery navigation trails extracted from search logs have been shown to exhibit traits of orienteering behavior (White and Drucker, 2007).

Another motivation behind the need for significant amounts of postquery interaction is that users may not understand the inversely proportional relationship between precision and recall. To obtain high precision, users seek a small result set comprising mainly of relevant documents. When targeting such precise result sets, users miss many other relevant items and therefore obtain low recall. To achieve high recall, users issue broad or general queries that retrieve large result sets. However, since these large sets contain more irrelevant information, precision is lowered. In attempting to balance precision and recall, it is unlikely that users will locate the information they require in a single result set, necessitating berrypicking-style behavior (Bates, 1984).

In Figure 2.2, we present a model of exploratory search behavior that illustrates the strong interaction between the search process and the problem context. It is assumed that there is some internal or external information-seeking context motivating the information problem and that the user has already recognized the need to address the problem (although the problem may not be well-defined initially). This activity may occur over multiple search sessions and an extended period of time. It may also repeat many times within a multifaceted exploratory task, as users may exhibit this behavior for each aspect of the task. Although Figure 2.2 shows focused search following exploratory browsing, it is also likely that during the search process, the exploration and focused search activities will alternate, depending on the user's perception of the information problem. For example, high levels of uncertainty or confusion may lead the user to return to browsing even though they once had a clear understanding of the problem context.

Figure 2.2 illustrates the constant interplay between the problem context and the search process during the course of an exploratory search. Time progresses from left to right. Initially, users are likely to explore the space and better define and understand their problem. As they explore, their perceptions of the problem may fluctuate dramatically. During this period, the problem context and the exploratory browsing behavior are highly dynamic. In this stage, the problem is limited, labeled, and a framework for the answer is defined. In exploratory searches, the answer framework may be more uncertain and require more definition over the course of the search. Over time, the problem becomes better defined, and the user is able to conduct targeted searches involving automatic search systems. It is worth noting that the problem can also become more confusing or challenging as the search progresses (e.g., the more we learn about the topic, especially if we are novices in the domain, the more confused/overwhelmed we may get). In the focused search phase, users (re)formulate query statements, examine search results, and extract and synthesize relevant information.

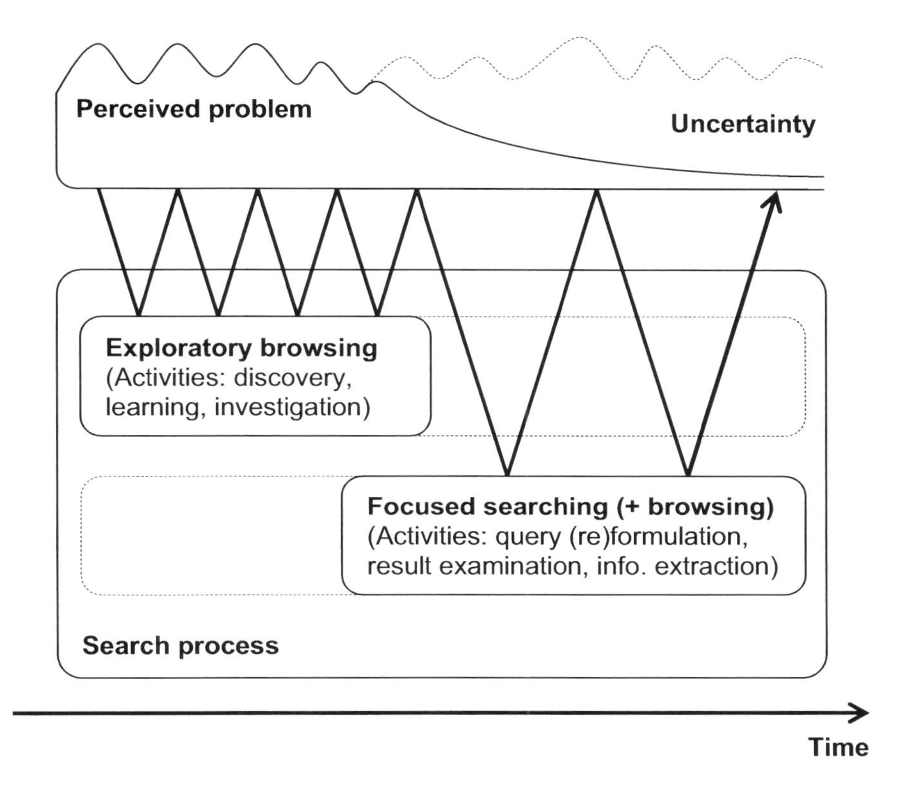

FIGURE 2.2: A model of exploratory search behavior.

The transition between exploratory browsing and focused searching (or vice versa) may result from a single "eureka" moment, but more likely a scaffolding process, where uncertainty is gradually reduced. There is likely some overlap between the two phases. For example, queries may be used early in the search to get the user near relevant documents, or orienteering may be employed to navigate the document space following focused queries. It is also possible for either phase to be removed completely. Exploratory browsing sessions are similar to exploratory data analysis, where the goal is to generate a set of hypotheses from data. Focused searching sessions may contain multiple query iterations and little postquery navigation, and are associated with poor retrieval system performance or tasks with multiple facets. In such situations, the user goals may be clear (i.e., there is low uncertainty), but they may need to issue many different queries or examine many sets of search results to resolve their information problem.

The elucidation of vague information needs and the resultant reduction in uncertainty is one of the defining characteristics of exploratory searches. Such interactions between the search process, its outcomes, and the problem context have been discussed in previous work, such as Dervin's sense-making model (1977), Wilson's (1997) model of information behavior, and Kuhlthau's information search process model (1991, 2004). Dervin investigated individual sense-making, developing theories underlying the "cognitive gap" that individuals experience when attempting to rationalize observed data. Wilson's model on information behavior suggests that intervening variables (e.g., psychological, demographic, social) may increase or decrease uncertainty during a search and that information seeking can only affect the user-in-context once outcomes are known. Kuhlthau's model of the information search process is typified by uncertainty in the early stages, which is reduced as the search proceeds. The reduction in the uncertainty of the problem situation results from changes in knowledge state; as users become more knowledgeable about the subject matter, they can construct well-formed query statements and need to browse less.

Exploratory search covers more than the iterative query refinement strategy discussed for decades in the research literature (e.g., Belkin et al., 2001) and supported by many existing search systems, they are inherently open-ended. Exploratory searches involve learning more about the topic of the search, understanding the nature of the document collection, and investigating browsing opportunities in real time as they occur during result examination. Figure 2.3 illustrates an analytical iterative query refinement strategy and an exploratory search strategy.

The exploratory and iterative search strategies differ in how the searcher traverses the information space and in knowledge of the destination. In the iterative search strategy, the target of the search is typically known, and the search task is to find the target. Iterative search strategies can involve the consultation of thesauri before search, the use of query suggestions ("related searches") offered by commercial search engines, and the systematic refinement of query statements (Marchionini and Shneiderman, 1988). In a recent user study of a search engine interface enhancement known as popular destinations (i.e., domains where the majority of users ended up after issuing a search query, as determined from search logs), White and colleagues (2007) found that query suggestions were generally most useful for refining the system's representation of the user's need, rather than initiating new directions for the searcher to investigate. Since the problem context may be open-ended or multifaceted in exploratory searches, a single target answer may not exist. In such situations, information novelty and information coverage are important aspects, as well as situational relevance (Saracevic, 2007).

In the exploratory search strategy, searchers visit more of the information space, and many search targets may be present, each coresponding to an aspect of the task. Within exploration, there may be some degree of progressive narrowing as part of the exploration-enrichment-exploration trade-off (Patterson et al., 2001). Search under this model begins with the retrieval of a broad set of documents, such as one retrieved by a high-recall/low-precision query, then proceeds with narrowing that set down into progressively smaller, higher-precision result sets, before reading the

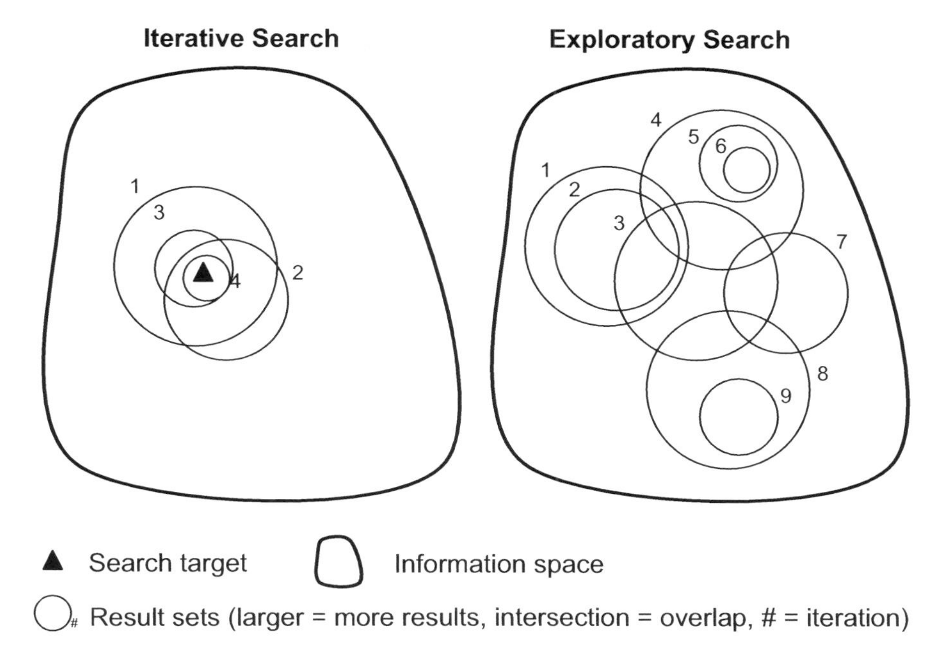

FIGURE 2.3: Iterative search versus exploratory search strategies.

documents and extracting the information. This behavior may be observed in Figure 2.3, between Exploratory Search query iterations 4–6.

Exploratory searches may also seek the discovery of gaps in existing knowledge so that new research ground can be forged or unpromising directions can be avoided. For example, Garfield (1970) proposed the notion of a "negative search," where the failure to retrieve results for a search query may actually be a positive outcome if the goal is to propose a new solution or a new problem, as is common practice in the scientific community.

2.4 DIFFERENTIATING EXPLORATORY SEARCH

The following are attributes of exploratory search that differentiate it from other types of information seeking and related disciplines:

1. Exploratory search sessions can transcend multiple query iterations and potentially multiple search sessions. An exploratory search can last for days, weeks, or months depending on the nature of the search task (e.g., a search for information pertaining to a digital cam-

era purchase may take less time than the set of searches related to a research paper). It is important that exploratory search systems support searches over time. Examples of this type of support include session memory features that store recent queries and long-term user histories that retain information on user preferences and searches over many search sessions.

2. The information need that motivates an exploratory search is generally open-ended, persistent, and multifaceted. Open-endedness relates to uncertainty over the information available, or incomplete information on the nature of the search task. Exploratory searches are based on a mixture of specific and diversive curiosities (Berlyne, 1960), and emphasize learning and investigation.
3. The goal of the search extends beyond simply locating information toward activities associated with learning and understanding. That is, the search task does not exist in isolation from the surrounding task context. Not only does the context influence the performance of the task, but it also influences what action should be taken with the found information. Exploratory searches are generally conducted to help people make more informed decisions or to improve their understanding of a topic. The emphasis on intelligence amplification and personal development is less evident in other types of information seeking such as information foraging or berrypicking.
4. The interaction behaviors observed during an exploratory search are generally a combination of browsing and focused searching, with more emphasis on the former. People use browsing as a way to resolve the uncertainty and confusion that can occur as new information is encountered.
5. Exploratory searches may involve the collaboration of multiple people in a synchronous or asynchronous manner. Given the strong relationship between exploratory search and information use and information understanding, it is likely that these searches will involve engagement with other people during the search. These people may be involved in the specification of the goals that drive the task (information need creators) and are therefore interested in the task outcomes (e.g., a manager). Also, people may be involved in the completion of the task (e.g., friends planning a vacation or coworkers working toward a shared goal).
6. The evaluation of systems to support exploratory search requires a methodology that targets learning and insights, as well as task outcomes and system utility. To determine how well systems support exploratory search activites, they must be evaluated in terms of their ability to facilitate the key elements of search exploration (e.g., helping users obtain new insights, assisting learning, offering support for critical decision making).

2.5 SUMMARY

In this chapter, we presented a definition for exploratory search. We targeted two important elements: the problem context that motivates exploratory searches and the search process characterizing exploratory search behavior. We also proposed a model of exploratory search that illustrates the interaction between the search process and the problem context. We differentiated exploratory search from other classes of information seeking. In Chapter 3, we compare and contrast this definition with related work in fields such as human–computer interaction, IR, information science, and psychology.

* * * *

CHAPTER 3

Related Work

Exploratory search is a multifaceted concept; it is constantly being changed and shaped by a range of related research. When defining what constitutes an exploratory search, aspects of research in a number of areas apply. In this chapter, we will offer relevant theories from related disciplines such as human–computer interaction, IR, information science, and psychology. We relate each body of work to exploratory search and highlight differences where appropriate.

3.1 CURIOSITY, EXPLORATORY BEHAVIOR, AND BROWSING

Curiosity is an emotion that may cause natural inquisitive behavior such as exploration, investigation, and learning, evident by observation in many animal and human species. Exploratory behavior, defined by the National Library of Medicine as "the tendency to explore or investigate a novel environment,"[1] is driven by curiosity and is evident in most exploratory searches. Both lookup and exploratory searches use curiosity in their search models, though different types.

Psychologist Daniel Berlyne (1960) proposed a categorization of different types of curiosity relevant to exploratory search: the distinction between specific and diversive curiosity. Specific curiosity is the desire for a particular piece of information, as typified by an attempt to solve a problem or puzzle. Diversive curiosity is a more general seeking of stimulation or novelty, as typified by a bored television viewer flipping between channels. In information seeking, specific curiosity corresponds to well-defined goals and directed searching, while diversive curiosity corresponds to ill-defined goals and exploratory browsing (Pace, 2004).

Switching between the two types of navigational behavior is necessary in many search tasks; however, it is not detailed in Figure 3.1. Other psychological aspects, such as anxiety from searching under pressure and time constraints, direct the strategies used by Web users.

Berlyne posited three stages of exploratory behavior: (1) orienting responses, (2) locomotor exploration, and (3) investigatory responses (1960). In terms of exploratory search, these parallel: (1) obtaining overviews of the data, perhaps through multidimensional information visualizations and surveying the information landscape for potential next steps, (2) focusing on a specific object,

[1] http://www.nlm.nih.gov/cgi/mesh/2008/MB_cgi.

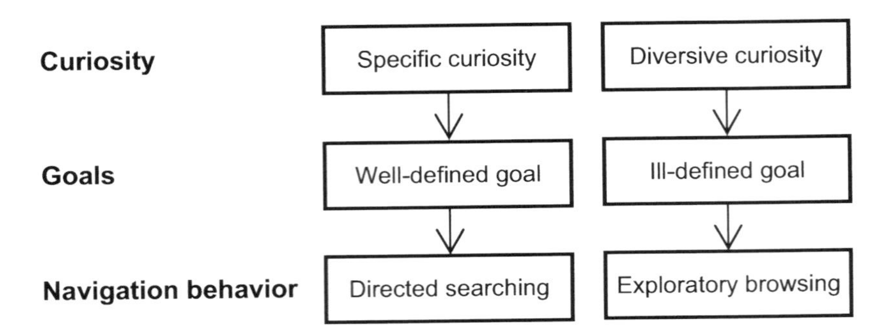

FIGURE 3.1: Variants of curiosity and their roles in search (based on Pace, 2004).

such as a potentially relevant or interesting document, and (3) examining that object in more detail. Wolfe's model of visual search (1994) is similar, and it postulates that one surveys the available information initially, then targets points of interest within the broader visual field for more complex interpretation and understanding.

In a recent review, Hughes (1997) traces the history of the various theories developed to explain exploratory behavior in psychology. Hughes defines intrinsic exploration as follows: "Intrinsic exploration involves exploratory acts that are not instrumental in achieving any particular goal other than performance of the acts themselves." This is contrasted with extrinsic search, driven by a goal, such as the need for food or escape from danger. Information foraging (Pirolli and Card, 1995, 1999), discussed in more detail in the next section, is an example of extrinsic search. In contrast, exploratory search is more closely related to intrinsic exploration, although there may be tasks or learning constraints that restrict it from being purely exploratory.

In an extensive review of curiosity, Loewenstein (1994) found available theories only partially explanatory. He posits a sense-making theory for explaining curiosity that is similar to work in information-seeking (e.g., Belkin, 1978; Dervin, 1977). His theory states that in the animal kingdom, motile animals' exposure to new environments, stimuli, or information bring the possibility of discovering new food sources, mates, nesting or sleeping sites, or ways to escape predation. In such settings, exploratory behavior is valuable and often underappreciated, although too much exploration can be dangerous (Bell, 1991). Excessive exploration in risky environments leads to death or failure to reproduce often enough, resulting in a decline in the species. Too static a pattern and the animal may lose out to competing species that explore and discover more items of value to their survival. Curiosity-driven behavioral patterns also apply to exploratory search, where the patterns are browsing interactions.

Bates (2004) suggests that browsing is a cognitive and behavioral expression of exploratory behavior and she claims that it has four elements: (1) glimpse a scene; (2) target an element of a scene visually and/or physically (if two or more elements are of interest, they are examined serially, not in parallel); (3) examine item (s) of interest; and (4) physically or conceptually acquire or abandon examined item(s). This sequence is repeated indefinitely as people explore in satisfaction of their curiosity. To this end, exploratory search systems should offer collection overviews (glimpses), the ability to traverse trails through the collection (exploratory browsing), and document examination/retention.

3.2 INFORMATION FORAGING

Information foraging theory (Pirolli and Card, 1995, 1999) attempts to explain information-seeking behavior in humans. Central to the theory is the idea that information foraging is similar to food foraging mechanisms in living organisms. Therefore, optimal foraging theory (Stephens and Krebs, 1986) helps researchers understand foraging behavior in human consumers of information (or "informavores" as they were referred to in Chapter 1). Optimal foraging models facilitate the investigation of foraging behavior in relation to particular environmental conditions in a dynamic ecology.

Optimal foraging theory contains a patch model and a diet model. The patch model addresses decisions related to searching and exploiting an environment that has a patchy distribution of resources. The conventional patch model typifies situations in which organisms face an environment where food is distributed in a patchy manner. Exploratory searches involve similar constraints where foragers spend time moving between information patches and acquiring knowledge rapidly when a relevant patch is encountered. The forager has the choice to remain in the current patch or attempt to find a new patch if the information resources are becoming exhausted.

The diet model specifies the types of prey to eat and that to ignore. When moving between patches, foragers may select the information prey to maximize the rate of gain of information relevant to their task. From such analysis, one can model information foraging theoretically in terms of cumulative gain or rate of gain over time, e.g., Charnov's marginal value theorem (1976). For more details on the relationship between optimal foraging and information foraging, see Pirolli (2008).

Users assess the appropriateness of a particular trail by considering a representation, usually a textual description such as a search engine result caption or a thumbnail image, of the distal content. Furnas (1997) suggested that a representational object held a "residue" of the information item it represented. The concept of residue was refined by Pirolli (1997) as information scent and defined in Card and colleagues (2001) as a user's "(imperfect) perception of the value, cost, or access path of information sources obtained from proximal cues, such as WWW links." In the initial work on information foraging, Pirolli and Card (1995, 1999) defined the profitability of an information source "as the value of information gained per unit cost of processing the source." Cost is defined in terms of time spent, resources utilized and opportunities lost when pursuing a search strategy instead of others (Russell et al., 1993).

Information foraging provides information-seeking researchers with a way to examine user goals, their decision-making processes, and adaptation to the information environment. Computational cognitive models of information foraging have been created to inform the design of information systems based on rational analysis (Anderson, 1990). Adaptive control of thought in information foraging (ACT-IF; Pirolli, 1997) models optimal foraging behavior in large text collections using the Scatter/Gather browser interface (Cutting et al., 1992). Research on inferring user needs by information scent (INUIS) shows that users can be clustered into types or profiles based on their surfing patterns (Chi et al., 2001; Heer and Chi, 2001). Collaborative filtering allows users to forage for information in groups much like a group of humans who band together to hunt for food when items included in their diet are distributed sparsely in their environment. Through ascribing a history of use to a digital object, a single user can benefit from the foraging of others (Wexelblat and Maes, 1999). Logs of users' click-through and other foraging behaviors are commonly used by Web search engines to improve search result ranking (e.g., Agichtein et al., 2006; Bilenko and White, 2008).

Information foraging and exploratory search are similar in a behavioral sense. Exploratory searchers navigate between information patches based on information scent and consume information relevant to their goals. However, information foraging is an extrinsic model that depends on the existence of a constrained information diet. This is similar to an animal having a small set of desired prey. However, in exploratory search, the prey may be unknown at the outset of the search and may change dramatically based on encountered information. As suggested in the previous section, exploratory searches may be motivated by curiosity rather than a need for information. The information need in such circumstances is not as pressing, and abandonment of the search may be an acceptable outcome. In addition, it may not be an exploratory searcher's goal to maximize their rate of information gain. During exploratory search episodes, users may use seemingly irrational search strategies (e.g., retracing their steps through the information space to revisit a branching point or reading redundant information to confirm a hypothesis) to help them acquire the knowledge they seek and understand the information they encounter.

3.3 BERRYPICKING

Bates (1989) developed the berrypicking approach to information-seeking behavior. The term "berrypicking" is an analogy to picking berries in a forest; berries are scattered on bushes, not in bunches. People must pick the berries singly. In a similar way to information foraging and wayfinding (Lynch, 1960), the approach views the searcher as moving through an information space, gathering fragments of information as they move and seeking cues that aid navigation decisions. However, the emphasis in berrypicking is on the dynamism of needs during the search, rather than the act of searching (foraging) itself.[2]

Berrypicking states that new information encountered gives the searcher new ideas and directions to follow and, consequently, a new conception of the query. At each stage of the search process, searchers are not just modifying the search terms but the query itself. Bates described the approach as an "evolving search"; as the search progresses, the desired outcome may also change. Evolving or berrypicking searches have been studied in recent years (e.g., Campbell, 1999; Ellis, 1984; Kuhlthau, 1993; Marchionini, 1995; Vakkari and Hakala, 2000; White and Drucker, 2007). As suggested earlier, users' perceptions of relevance are also likely to change during the course of a search (Harter, 1992; Saracevic, 1997, 2007; Spink et al., 1998; Tang and Solomon, 1998).

At each stage of the search, with each different conception of the query, the user may identify useful information and references. The query is not satisfied by a single final retrieved set, but by a series of selections of individual references and fragments of information at each stage of the ever-modifying search. Searchers' understanding of their information need is enhanced as they encounter additional information during a search. Campbell (2000) suggested that this enhancement occurs to support or deny beliefs in various aspects of the need. The searcher revises their beliefs in what information is relevant until it reaches an end point of redundancy. Redundancy may arise because the information need has been satisfied, or it no longer has perceived importance to the searcher. Figure 3.2 shows an example of an evolving, berrypicking search.

Berrypicking is a commonly used strategy in exploratory searches, and Figure 3.2 is a representative form of exploratory search behavior. During exploratory searches, the evolution of the information need is important, and a core activity is gathering and understanding information fragments.

The focus of the classic model presented in Figure 1.1 is query-document matching. In contrast, the focus of the model in Figure 3.2 is the sequence of searcher behaviors. The line of the arrow represents the continuity of a single human moving through many actions toward a general goal of a satisfactory completion of research related to an information need. The search evolves as the individual follows various leads and shifts in thinking, illustrated by changes in direction.

In the case of classic lookup-based search model illustrated in Figure 1.1, the line would be short and straight, with a single query and a single set of information items output. The berrypicking model differs from the lookup model in two respects: (1) The nature of the query is an evolving

[2] Note that berrypicking of information without the search need itself changing (evolving) is possible (Bates, 1989), but the evolution is our key differentiator with information foraging.

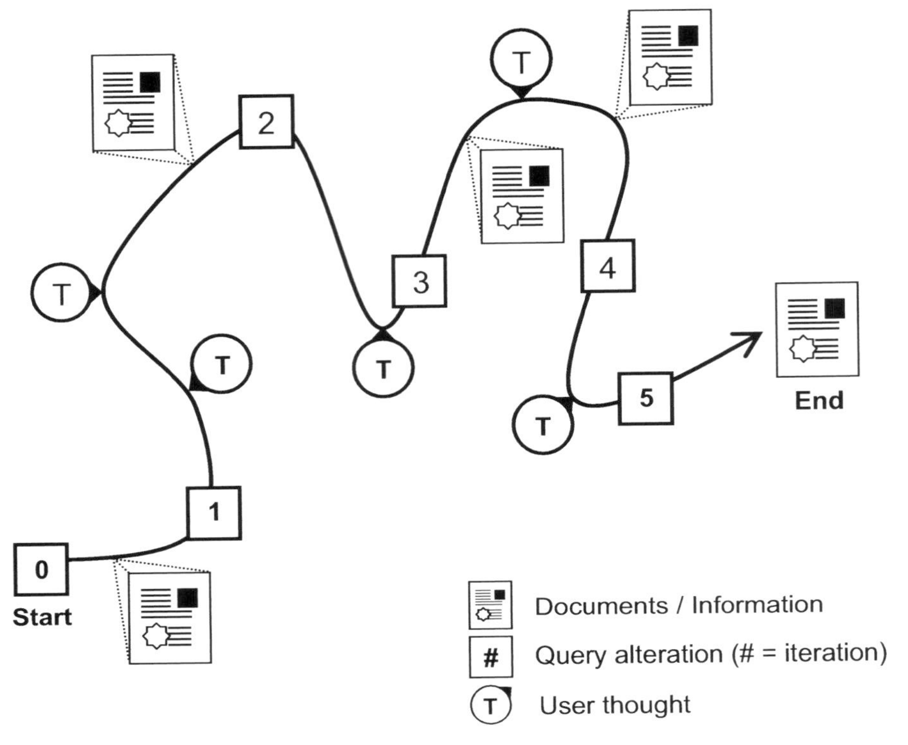

FIGURE 3.2: An evolving berrypicking search (based on Bates, 1989).

one, rather than solitary and unchanging, and (2) the nature of the search process follows a berrypicking pattern, instead of leading to a single best retrieved set.

Within berrypicking, a variety of search activities can take place. Bates (1990) described a four-level hierarchy of search activities: move, tactic, stratagem, and strategy. Moves are single actions performed by the user, either physically or mentally; examples of mental actions are deciding or reading. Tactics are a combination of moves. There are numerous combinations of moves that can be used to support a tactic. Stratagems are a larger combination of moves and tactics. Marchionini (2006a) noted a series of exploratory search activities that could be considered as stratagems (e.g., comparison, discovery, synthesis). Strategies involve a combination of moves, tactics, and stratagems. Strategies are heavily dependent on the current task context, such as finding pertinent research for a journal article. Berrypicking could be considered a complex combination of tactics and moves, whereas a simple lookup could be a simple set of tactics and moves.

In berrypicking, the information viewed by the searcher is typically used to inform subsequent moves and tactics, such as the queries to issue or documents to examine (as shown in Figure 3.2). Although this session-related learning may also occur during exploratory searches, there is greater likelihood that during exploratory searches, encountered knowledge will be transformational, and it will create significant changes in terms of new stratagems and strategies. Searchers may decide to search within a new domain, use a different search system or interface, or adopt a different search strategy, perhaps involving collaboration with others through email, telephone, or in-person meeting.

3.4 SENSE-MAKING

Sense-making is the creation of situational awareness and understanding in situations of high complexity or uncertainty in order to make decisions. It is "a motivated, continuous effort to understand connections (which can be among people, places, and events) in order to anticipate their trajectories and act effectively" (Klein et al., 2006a). People may engage in sense-making tasks frequently; however, exploratory searches always involve some degree of sense-making. Sense-making generally involves the following steps (Stefik et al., 1999; Pirolli and Card, 2005) identified through observation and cognitive task analysis: (1) knowledge gap recognition; (2) generation of an initial structure or model of the knowledge needed to complete the task—concepts, relationships, and hypotheses; (3) search for information; (4) analysis and synthesis of information to create insight and understanding; and (5) creation of a knowledge product or direct action based on the insight or understanding.

Sense-making typically involves a series of continuing gap-defining and gap-bridging activities between situations (Dervin, 1992, 1998). It is an active two-way process of fitting data into a frame (mental model) and fitting a frame around the data. Neither data nor frame comes first; data evoke frames, and frames select and connect data. When there is no adequate fit, the data may be reconsidered, or an existing frame may be revised (Klein et al., 2006b).

Research in cognition, learning, and task-based information seeking and use provides important insights for understanding sense-making. Researchers have proposed several models to capture the processes involved in sense-making (Dervin, 1992, 1998; Dervin and Nilan, 1986; Pirolli and Card, 2005; Qu and Furnas, 2007; Russell et al., 1993).

Dervin and colleagues (1986, 1992, 1998) focus on developing sense-making theories underlying the "cognitive gap" that individuals experience when attempting to make sense of observed data (e.g., Dervin and Nilan, 1986). In the model, users which have a particular task and situation encounter a trouble spot or a "gap" that impedes their progress. The user, or human actor, must overcome the gap by finding help or making sense of the current situation in order to attain their desired outcome. In this model, Dervin and Nilan regard information seeking as a situation-sensitive sense-making process. They suggest that information seeking should be holistic and focus on: subjective information constructed by human actors; constructive actors with internal (or cognitive) conceptions as opposed to passive receivers of information; and situations within which actors act (including preceding and following information system use) (Ingwersen and Järvelin, 2005). In a similar way to sense-making, the anomalous state of knowledge (ASK) hypothesis states that there is a gap between what one knows and what they would like to know, and the need to fill the gap is what drives one to seek and retrieve information. The ASK hypothesis, proposed by Belkin and colleagues (1978, 1982a,b), is an important aspect of information seeking because it identifies the motivation behind it.

Russell and colleagues (1993) described sense-making as "the process of encoding retrieved information to answer task-specific questions." They defined a sense-making model comprising four main processes: (1) search for representation (structure): the sense-maker creates representations to capture salient patterns of data; (2) create instances of representations: the sense-maker identifies information of interest and encodes it in the representation; (3) modify representation: representations are modified during sense-making when data is ill-fitted or missing in the representation; and (4) consume instantiated representations: the sense-maker consumes the instantiated representation and uses it in performing the task. Russell's model indicates the iterative nature of sense-making. The processes may be followed for several iterations until the sense-making is successful. Much like exploratory search, sense-making is therefore an iterative process that occurs on exposure to information driven by a desire to understand and use that information. Qu and Furnas (2007) separate the search for structures from the search for data in the sense-making process. They also integrate the two processes and emphasize the bidirectional relationship between search and representation construction (Zhang et al., 2008).

Pirolli and Card (2005) conducted a cognitive task analysis and identified two loops of sense-making activities: (1) an information foraging loop that involves searching for information, filtering it, and reading and extracting information into some schema, and (2) a sense-making loop that involves iterative development of a mental model (a conceptualization) from the schema that best fits the evidence. A variety of conceptual changes can happen to the mental representation of knowledge as a sense-maker learns about the task, problem, or situation. Researchers have identified various degrees of change, ranging on a continuum from the addition of facts, weak revision, or radical restructuring (Vosniadou and Brewer, 1987). Piaget (1978) recognized two types of conceptual change in knowledge acquisition: (1) assimilation: the addition of information to existing knowledge structures, and (2) accommodation: the modification or change of existing knowledge structures.

Information processing is driven by inductive processes (from data to theory) or structure (from theory to data). The foraging loop is a trade-off among three kinds of processes: information exploration, information enrichment, and information exploitation (e.g., reading). Typically, users cannot explore all documents, and must forego coverage in order to actually enrich and exploit the information. The sense-making loop involves substantial problem structuring (the generation, exploration, and management of hypotheses), evidentiary reasoning (marshalling evidence to support or disconfirm hypotheses), and decision making (choosing a prediction or course of action from the set of alternatives). These processes are affected by many well-known cognitive limitations and biases (Pirolli, 2009).

Among several task characteristics recognized by Kim and Soergel (2005), the tasks that require at least some degree of sense-making often involve: (1) new situations or problems; (2) complex, less structured situations or problems; (3) a new domain; and (4) an unclear information need. There is an overlap between such situations and the definition of exploratory search offered in Chapter 2. Exploratory searchers are constantly engaged in sense-making activities as they move through the information space. These movements are interrupted when a gap is encountered that requires information to be bridged. Sense-making is an individual process of construction, not a process of utilizing existing information. Exploratory searches typically involve a prolonged engagement in which individuals iteratively look up and learn new concepts and facts. The knowledge acquisition causes the searcher to dynamically change and refine their information goals, and to ask more informed questions that probe deeper into the problem and the information space. Exploratory search can be viewed as a subcomponent of sense-making.

3.5 INFORMATION-SEEKING PROCESSES AND BEHAVIORS

Exploratory search is a class of information seeking. As mentioned earlier, a number of models of the information-seeking process and information-seeking behavior have been developed (e.g., Choo et al., 2000; Ellis, 1989; Kuhlthau, 1991; Marchionini, 1995; Wilson, 1997).

The Ellis Feature Set is a set of eight features that form a framework for information-seeking behavior (Ellis, 1989; Ellis et al., 1993; Ellis and Haugan, 1997). The feature set differentiates the various information-seeking patterns of scientists and engineers in their individual surroundings. The features are listed below (Ellis and Haugan, 1997): (1) starting: activities such as the initial search for and overview of the literature or locating key people working in the field; (2) chaining: following footnotes and citations in known material or "forward" chaining from known items through citation indexes or proceeding in personal networks; (3) browsing: variably directed and structured scanning of primary and secondary sources; (4) differentiating: using known differences in information sources as a way of filtering the amount of information obtained; (5) monitoring: regularly following developments in a field through particular formal and informal channels and sources, (6) extracting: selectively identifying relevant material in an information source; (7) verifying: checking the accuracy of information; and (8) finding: activities actually finishing the information seeking process.

Most situations involving information seeking can be characterized by the Ellis model. However, the model does not capture the main aspects of an exploratory search process. The feature set excludes external causative factors, and an individual is not guaranteed to undergo an identical information-seeking process as outlined in the model. In addition, the model does not support tasks or IR, and it is unidirectional; it does not analyze relationships among the features.

Choo and colleagues (2000) developed a model of online information seeking that combines both browsing and searching. It suggests that much of Ellis's model is already implemented by components currently available in Web browsers. Searchers can begin at a Web site (starting), follow links to information resources (chaining), bookmark pages (differentiating), subscribe to services that provide electronic mail alerts (monitoring), and search for information within sites or information sources (extracting).

Kuhlthau (1991, 1993) developed a model that identifies and emphasizes the importance of the individual stages that learning tasks and problem solving involve. Kuhlthau (1991) proposed that the feelings of doubt, anxiety, and frustration are natural and play a role in information seeking. Occurrence of these feelings has already been studied (Ford, 1980; Mellon, 1986), and anxiety has usually been associated with a lack of knowledge of information sources and apparatus. Information seeking, by its very nature, causes anxiety because there is no guaranteed positive outcome to the search (i.e., the searcher can be unsuccessful in finding what they seek).

Kuhlthau's research was completed using a series of longitudinal empirical studies conducted on students and library users. Kuhlthau's information-search process model highlights the differences in feelings, thoughts, and actions that people experience during the search process. Changes in feelings, thoughts, and actions of the user are stage dependent, and each task is unique to the stage of the investigational process. Kuhlthau's model, followed by a detailed description of the stages involved, is given in Table 3.1.

Kuhlthau's stages, as interpreted by Ingwersen and Järvelin (2005), are as follows: (1) initiation: becoming aware of the need for information, when facing a problem; (2) selection: the general topic for seeking information is identified and selected; (3) exploration: seeking and investigating information on the general topic; (4) focus formulation: fixing and structuring of the problem to be solved; (5) collection: gathering pertinent information for the focused topic; and (6) presentation: completing seeking, reporting, and using the result of the task.

Kuhlthau's model is unique to her predecessors in its incorporation of the psychological aspect of search into information seeking, concurrent with today's interpretation of exploratory search. The notion of exploration is fundamental to exploratory search, and Kuhlthau's model outlines exploration as one of the primary six tasks that the user carries out during search. Exploration, as used in her model, is defined as being an investigational stage of the information-seeking process. In the model, the actions of the user/actor transition from exploring to documenting. Exploratory search differs from Kuhlthau's ideals because it consists primarily of exploration, instead of exploration being a fraction of the entire search process.

TABLE 3.1: Kuhlthau's information search process (2004). Copyright © Libraries Unlimited 2004. Used with permission.

TASKS	INITIATION	SELECTION	EXPLORATION	FORMULATION	COLLECTION	PRESENTATION
Feelings (affective)	Uncertainty	Optimism	Confusion, frustration, doubt	Clarity	Sense of direction/ confidence	Satisfaction or disappointment
Thoughts (cognitive)		Vague	————————→		Focused	
			————————→		Increased interest	
Actions (physical)	Seeking relevant information, exploring		————————→		Seeking pertinent information, documenting	

Marchionini (1995) proposes another model of the information-seeking process, directed toward electronic environments. In his model, the information seeking process is composed of eight subprocesses which develop in parallel: (1) recognize and accept an information problem, (2) define and understand the problem, (3) choose a search system, (4) formulate a query statement, (5) execute search, (6) examine results, (7) extract information, and (8) reflect/iterate/stop. This model defines the activities at each stage and is more suitable for electronic environments than the Ellis model. Figure 3.3 illustrates Marchionini's model with transitions between each of the eight stages highlighted.

The information-seeking process model captures many important elements of information seeking, including aspects of collection exploration in examine results, and aspects of knowledge acquisition in extract information. However, it does not have the same emphasis on learning and understanding as exploratory search nor does it fully represent search context or information use.

Wilson (1997) proposed that fields outside of information science, which include decision-making, psychology, innovation, health communication, and consumer research, are vital to the advancement of information behavior analysis. He advised information scientists to expand the scope of their research to include more disciplines. Indeed, as we demonstrate in this lecture, the need for interdisciplinary collaborating is also pressing in exploratory search. Wilson's model is a broad, static model which summarizes general information behavior, and it is not directly based upon empirical findings.

Wilson's model contains several elements that are valid when modeling exploratory search. His model includes a "person in context" or a person with a particular task at hand for which they

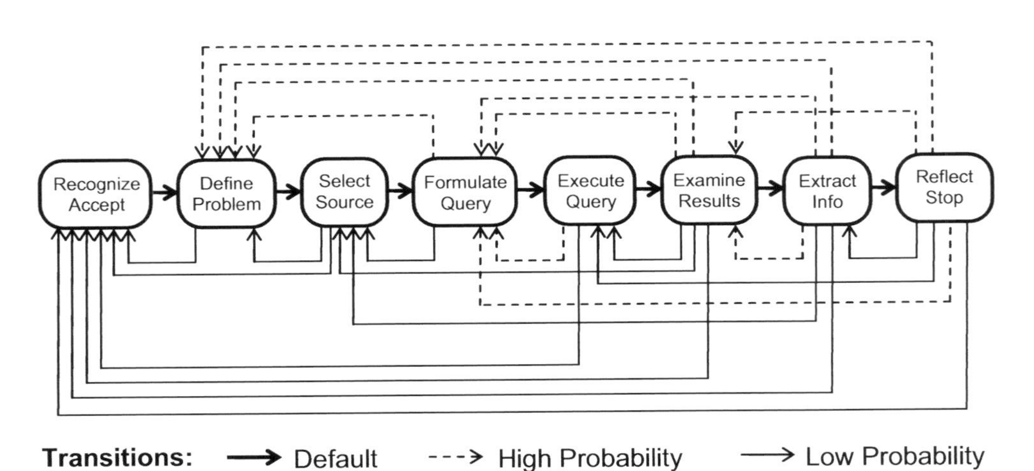

FIGURE 3.3: Information-seeking process model (based on Marchionini, 1995).

require information. His model shows that information-seeking behavior influences the person in context and their informational needs, which is also true of exploratory search. In other words, as a person searches, they may decide to investigate information other than that which they were initially seeking. In exploratory search, the search process profoundly influences users' task perceptions. Wilson's model illustrates that intervening variables (e.g., cognitive abilities, demographics, task and related environmental constraints) affect information-seeking behavior, which is true of exploratory search to a greater extent, given the focus on intelligence amplification.

3.6 COGNITIVE INFORMATION RETRIEVAL

Cognitive IR focuses on individuals' complex psychological functions during the retrieval process. Work in this area is relevant to exploratory search given the important role of cognition in learning and understanding. Following the development of a number of process models for IR (Belkin and Vickery, 1985; Henry et al., 1980; Saracevic et al., 1988), Ingwersen and Wormell (1989) devised a model, derived from Ingwersen and Pejtersen (1986). The model examined IR interaction and served as a forerunner for Ingwersen's cognitive model of information transfer (1992, 1996). Their model integrated both the systems-oriented IR research and cognitive IR research. Systems-oriented research includes authors' texts, text representation, IR techniques and queries. Cognitive IR research includes user's problem space, information problems, requests, interaction with intermediaries, and interface design. To further understand cognitive IR research, Wilson (1999) devised a model which summarized user-oriented (cognitive) IR research.

Wilson's model incorporates the concept of IIR under the heading of "information searching." The primary purpose of Wilson's model is to demonstrate the locality of the various elements of IR; in particular, the way in which they are allocated in a "nesting" within one another. Wilson's model places information search, or IIR, in the innermost part of the model, followed by information-seeking behavior and information behavior, respectively. The model decreases in specificity as it moves outward; information-seeking behavior makes up a portion of information behavior, and IIR is a specific type of information-seeking behavior.

In addition to Wilson's model of IIR, Ingwersen (1992) contributed to the understanding of IR interaction with his own conceptual model. Ingwersen's model incorporates the context, or the socio-organizational environment, of the information seeker. To further elaborate, context includes the scientific or professional domains with information preferences, and the strategies and work tasks that shape the user's awareness. In 1996, Ingwersen changed his model to include IIR, adding the work task and corresponding situation, as perceived by the user. The model emphasizes the primary elements of IR theory and the cognitive variation at any specific moment in time (found in documents, search engines, and in a user's cognitive space). Ingwersen's model not only looks at particular instances of time, but it also demonstrates the influence of context on information and system longitudinally. Longitudinal relevance also applies to Ingwersen's model in the social interaction between the work tasks and the user.

Saracevic (1996) devised a model which also examined situational relevance, while delving beyond to examine many types of relevance that are involved in IIR. Saracevic's model defines three different communication levels in IR interaction. One of the stratums is based on query and represents the data processing that occurs between source and interface. Another level of communication focuses on human computer interaction as it pertains to the need for information. Thirdly, Saracevic outlined the situational stratum, referring to information use in regard to the perceived work task, in the context of the environment. The Stratified Model is important because it highlights the adaptation that occurs from the system and searcher during IIR.

Ingwersen (1992, 1994) and Pao (1993) developed the principle of polyrepresentation, whereby cognitive overlaps occur during IR by different information structures (i.e., indexers vs. citations). Polyrepresentation can be used as a means for precise IR and for the expansion of intellectual availability of subject matter. Ingwersen (1996, 2001, 2002) expanded the concept of polyrepresentation upon examining the cognitive theory aspect of IIR. The five major information structures that embody polyrepresentation include: citations, author(s), indexers, selector(s), and thesauri. All of these structures have some degree of cognitive overlap, which is important to recognize when considering search techniques. This overlap may also be useful for exploratory searchers, who may seek highly reliable information sources or wish to use the overlap as a way to broaden their topic knowledge.

3.7 POSITIONING EXPLORATORY SEARCH

It is important when establishing exploratory search as a viable subdiscipline of information-seeking that we position it relative to existing disciplines such as IR, information visualization, information foraging, and sense-making. In this chapter and preceding chapters, we have described many of these disciplines and highlighted their relationship with exploratory search. In Figure 3.4, we present a Venn diagram that positions exploratory search in relation to other relevant areas. The figure illustrates the overlap between the research foci of the disciplines. For example, information foraging overlaps significantly with sense-making. However, information foraging covers activities and value functions (e.g., optimizing the rate of information gain) not fully represented in sense-making. In contrast, sense-making addresses issues of user comprehension and information use not generally present in information foraging models.

Exploratory search is a type of information seeking and a type of sense-making focused on the gathering and use of information to foster intellectual development. Overlap exists with a number of aspects of information seeking that are essential in exploratory search activities: (1) information visualization is an important element in hypothesis and insight generation, and for learning

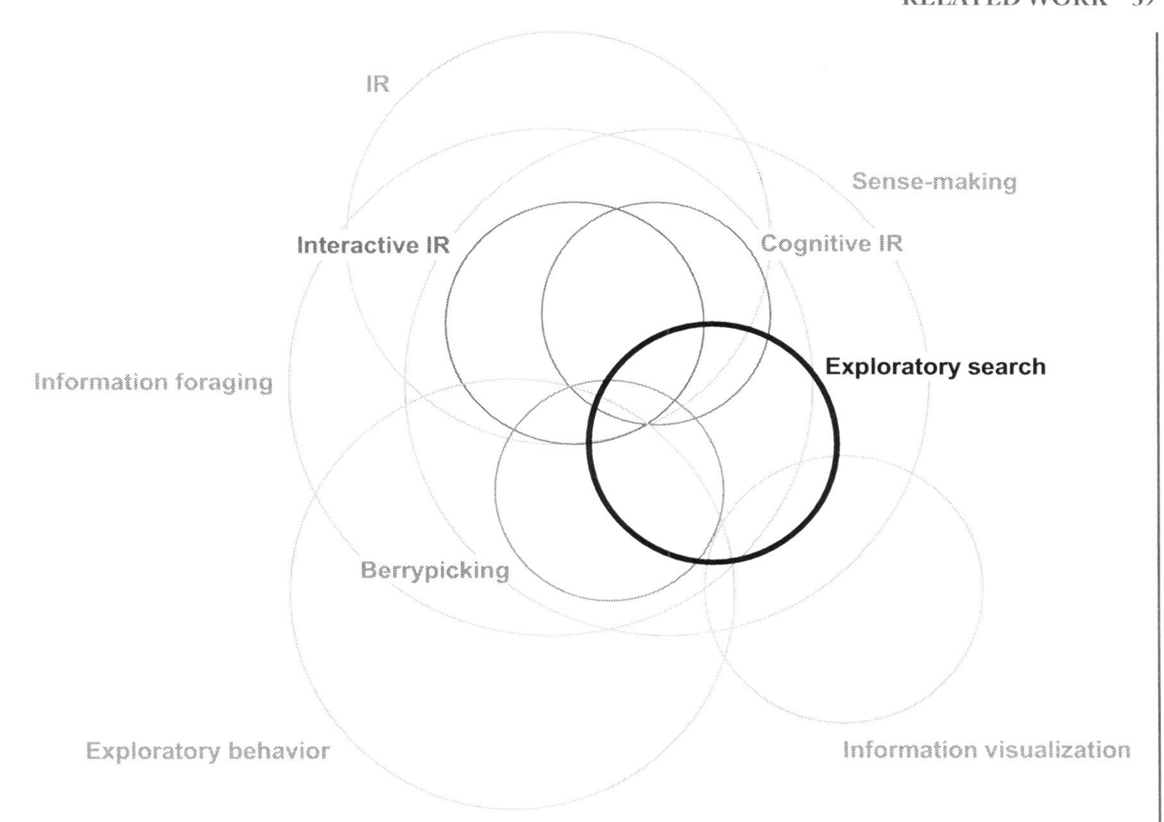

FIGURE 3.4: Venn diagram positioning exploratory search relative to other related research disciplines. Circle size signifies approximate size of each discipline. Color is used to differentiate interior circles.

about the information landscape; (2) exploratory behavior (browsing) is important in surveying and effectively navigating this landscape; (3) berrypicking and information foraging describe how users find information and adapt to their information environment; and (4) IIR and cognitive IR describe the behavioral and mental processes involved in finding information once the target is known. System designers can draw from research in all of these disciplines to better support and understand exploratory search behaviors.

3.8 SUMMARY

There is a wealth of research relevant to exploratory search. Such work gives us a solid foundation on which to build this emerging subdiscipline. Many of the theories described in this chapter are based on the findings of small user studies, potentially over long time periods. Given the recent availability of usage data from Web search engines and Web browser toolbars, it is possible to analyze the exploratory search behavior of hundreds of thousands of users (e.g., White and Morris, 2007), allowing targeted opportunities and services to be provided (e.g., alerting services could be created based on automatically generated user profiles and executed periodically). Tomorrow's models of search behavior will emerge from analysis of vast repositories of interaction data aggregated across many users. Although privacy concerns with the usage of such data will need to be resolved, there is an outstanding opportunity for system designers to draw upon existing exploratory search behavior to develop the exploratory search systems of the future.

In Chapter 4, we present a set of features that users can expect to see in exploratory search systems. We ground each feature suggestion in prior work and describe how its implementation will assist exploratory searchers.

• • • •

CHAPTER 4

Features of Exploratory Search Systems

The design of exploratory search systems (ESSs) presents unique demands, unlike designing for searches where the target is well known or where a single document or fact will suffice. Systems such as the mSpace Explorer (schraefel et al., 2005), the Relation Browser (Marchionini and Brunk, 2003), and Phlat (Cutrell et al., 2006) make search more effective by providing a broader range of interface functionality and dynamically updating presentation of search results in real-time during the session. Other options include the use of interfaces employing categorization or clustering (Hearst, 2006; Kules and Shneiderman, 2008). The development of new search tools requires novel research and collaborative efforts among computer scientists, social scientists, psychologists, library and information scientists, and practitioners who may lead the way with novel search applications on the Web. The provision of tools to support the exploration of such information spaces can yield great rewards for users, especially when contextual factors such as user emotion, task constraints, and dynamism of information needs are considered.

In this chapter, we propose a set of features that must be present in systems that support exploratory search activities. These features are summarized in the list below, along with a brief explanation of why they are appropriate and necessary for exploratory search systems:

1. Support querying and rapid query refinement: Systems must help users formulate queries and adjust queries and views on search results in real time.
2. Offer facets and metadata-based result filtering: Systems must allow users to explore and filter results through the selection of facets and document metadata.
3. Leverage search context: Systems must leverage available information about their user, their situation, and the current exploratory search task.
4. Offer visualizations to support insight and decision making: Systems must present customizable visual representations of the collection being explored to support hypothesis generation and trend spotting.
5. Support learning and understanding: Systems must help users acquire both knowledge and skills by presenting information in ways amenable to learning given the user's current knowledge/skill level.

6. Facilitate collaboration: Systems must facilitate synchronous and asynchronous collaboration between users in support of task division and knowledge sharing.
7. Offer histories, workspaces, and progress updates: Systems must allow users to backtrack quickly, store and manipulate useful information fragments, and provide updates of progress toward an information goal.
8. Support task management: Systems must allow users to store, retrieve, and share search tasks in support of multisession and multiuser exploratory search scenarios.

This list is partially derived from discussions between experts at ACM, NSF, and independent workshops organized on the topic of exploratory search systems. Users can expect to see an increasing number of systems offer these features as exploratory search support becomes more prevalent. We now describe each of these features in more detail, and where appropriate make reference to existing systems that support them.

4.1 SUPPORT QUERYING AND RAPID QUERY REFINEMENT

ESSs must offer their users the ability to specify information needs as search queries and refine those queries during the search session. Information needs can be expressed in the form of keyword queries or as natural language statements such as fully formed questions. Keyword queries are common in commercial Web search engines such as Google, Yahoo!, and Live Search, while free text is supported by search engines such as Ask.com.

Queries are used by the ESS to retrieve a set of information objects (documents, Web pages, fragments) presented to users in descending order of relevance. Support for queries and their subsequent refinement allows users to navigate to potentially relevant parts of the information space. This offers a shortcut to users, allowing them to browse rapidly to different parts of the space (or home in on a particular information target). However, user-defined queries may be based on users' existing knowledge and create limited opportunity for exploratory search. The presentation of query suggestions may help users select additional query terms. Query suggestions resulted from extensive work in the IR community on query expansion (see Efthimiadis, 1996).

Techniques such as RF (cf. Salton and Buckley, 1990) in the IR community and query-by-example (Zloof, 1975) in the database community can help users choose additional query terms. The techniques work by users providing the system with examples of relevant documents, and in turn, the system presents a set of related queries or documents. Large Web search engine companies can use their historical query log data to find queries commonly issued by other users immediately following the current query, and offer them as query suggestions to other Web searchers (e.g., Jones et al., 2006). Such suggestions are generally of most use for narrowing a search to target a particular subtopic (e.g., from [Hubble telescope] to [Hubble telescope pictures]) rather than supporting exploration (which may lead users to learn more about the telescope, for example). The reason for narrow suggestions is that most searches on the Web are for known items or are revisitations to previously encountered Web pages (Teevan et al., 2007). Many Web search engines now offer query completion drop-down menus that appear below the search boxes in their Web sites when users begin typing a search query. These are meant to give users support during the initial stages of their search (when they may be unsure of how to specify their needs) or to provide a shortcut to a query if they know exactly what to type. Few studies have been published on the effectiveness of real-time query completion techniques; although White and Marchionini (2007) did show that presenting candidate query expansion terms in real-time, as users typed their queries, was useful during the early stages of the search. White and Marchionini also demonstrated that the technique had the potential to lead to query drift if the suggested terms were not fully understood by the user.

Real-time query formulation support is one way systems can help users construct effective queries. Another way in which this may occur is through the use of dynamic queries (Ahlberg et al., 1992). Dynamic queries allow users to see an overview of the database, rapidly explore and conveniently filter out unwanted information. Users move through information spaces by incrementally adjusting a query (with sliders, buttons, and other filters) while continuously viewing changing results. Dynamic query interfaces use mouse actions such as slider adjustments and brushing techniques to pose queries and client-side processing, and immediately update displays to engage information seekers in the search process. Figure 4.1 shows an example of FilmFinder (Ahlberg and

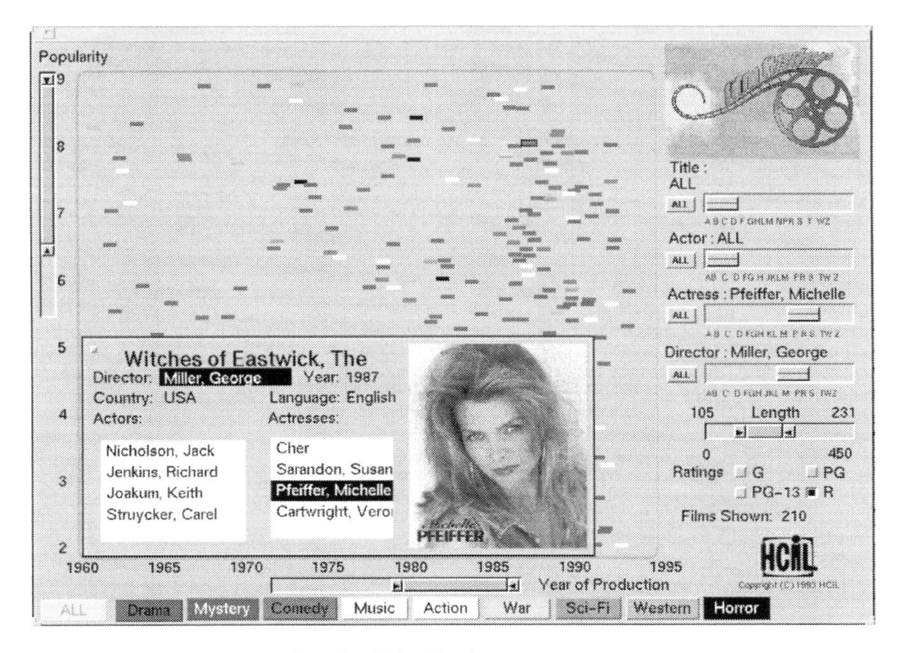

FIGURE 4.1: Dynamic query interface for FilmFinder.

Shneiderman, 1994), an early demonstration of the power of dynamic queries. As users move the slider bars, the points on two-dimensional scatter plot appear and disappear based on the slider values. Searchers can immediately observe the effects of their actions.

The tight coupling between queries and results [also observed in other techniques such as TileBars (Hearst, 1995) and Magic Lenses (Bier et al., 1993)] is especially valuable for exploration, where high-level overviews of collections and rapid previews of objects help people understand data structures and infer relationships among concepts. The ability to rapidly manipulate the data on a number of dimensions simultaneously is critical to hypothesis generation and information need clarification that occur during exploratory searches. Once users have concrete hypotheses and information needs, they then examine collection content in more detail to resolve their information problems.

4.2 OFFER FACETS AND METADATA-BASED RESULT FILTERING

Search systems must offer the ability for searchers to filter result sets by specifying one or more desired attributes of the search results. Information seekers often express a desire for interfaces that organize search results into meaningful groups, in order to help make sense of the results, and decide on actions (Hearst, 2006). Kules and Shneiderman (2008) proposed the use of categorized overviews of Web search results and showed that users were willing to change their search tactics to accommodate the overviews and improved their search effectiveness as a result. Two methods have been popular for generating useful document groupings: clustering and faceted categorization.

Clustering refers to the grouping of items according to some measure of similarity. In document clustering, similarity is commonly computed using associations and commonalities among features, where features are typically words and phrases (Cutting et al., 1992). The clustering procedure is fully automated, can be easily applied to any document collection, and can reveal interesting and unexpected trends in a group of documents (Hearst, 2006). Clustering can be useful for clarifying a vague query, by showing users the dominant themes of the returned results (Käki, 2005). It is also effective for query disambiguation, particularly for acronyms (e.g., the query "MSG" will return results for Madison Square Garden in New York City, and the food additive monosodium glutamate). Another aspect of clusters is their utility for eliminating groups of documents from consideration.[1] This result is supported by participant comments found in several studies (Käki, 2005; Kleiboemer et al., 1996). Hearst and Pederson (1996), and others (for example, Zamir and Etzioni, 1999) have used clustering of search results to make search more interactive.

[1] An alternative way to eliminate unwanted information from search results is through *negative RF* (Cool et al., 1996). This allows users to explicitly communicate which documents do not match their needs and exclude them and those similar from future result sets.

A facet is a "clearly defined, mutually exclusive, and collectively exhaustive aspects, properties, or characteristics of a class or specific subject" (Taylor, 1992). A keyword search brings together a list of ranked documents that match those search terms, whereas the goal of a faceted search is to enable a person to explore a domain via its attributes. Faceted interfaces aim to allow flexible navigation, provide previews of next steps, organize results in a meaningful way, and support both the expansion and refinement of the search.

Faceted categories are a set of meaningful labels organized in such a way as to reflect the concepts relevant to a domain (Hearst, 2006). They are usually created manually, although assignment of documents to categories can be automated to a certain degree of accuracy. Faceted search interfaces seamlessly combine keyword search and browsing, allowing people to quickly and flexibly find information based on what they remember about the information they seek. Faceted search interfaces can help people avoid feelings of being lost in the collection and make it easier for users to explore. This is an attractive feature of "view-based search systems" such as HiBrowse (Pollitt et al., 1994) that is accomplished by giving them a sense of the nature of the collection contents that is similar to browsing the shelves in a library or a supermarket. Flamenco (Yee et al., 2003) is a set of interfaces providing hierarchical, faceted metadata as entry points for exploration and selection. Figure 4.2 shows a screenshot of Flamenco in action for images in the Thinker collection of the Fine Arts Museum of San Francisco, with facet values "Asia" and "fabrics" specified.

The faceted interface presentation style gives users the opportunity to evaluate and manipulate the result set, usually to narrow its scope. It allows flexible ways to access the collection contents. Navigating within the hierarchy builds up a complex query over subhierarchies. The approach reduces mental work by promoting recognition over recall and suggesting logical but perhaps unexpected alternatives, while avoiding empty results sets. Meaningful categories support learning, reflection, discovery, and information finding (Kwasnik, 1999; Soergel, 1999). A drawback of these interfaces is the need for the manual creation of category hierarchies, although there has been some progress in the problem of semisupervised creation of faceted categories (Stoica and Hearst, 2004). Another drawback is that facets impose structure on the information space that may constrain free-form exploration.

mSpace (schraefel et al., 2006) offers a multicolumn-faceted spatial browser that presents persistent contextual information around items of interest. It improves access to information by supporting multiple ways of exploring the information itself. The mSpace model conceptualizes information as a set of "slices," and lets users "slice and dice" the information (Marchionini, 1995), in support of exploration.

Slices are arranged from left to right, in columns, creating a hierarchy, where the leftmost column is the top level of the hierarchy, and the rightmost is at the bottom. Items associated with a dimension are populated into a column. Figure 4.3 shows a screenshot of mSpace. The figure

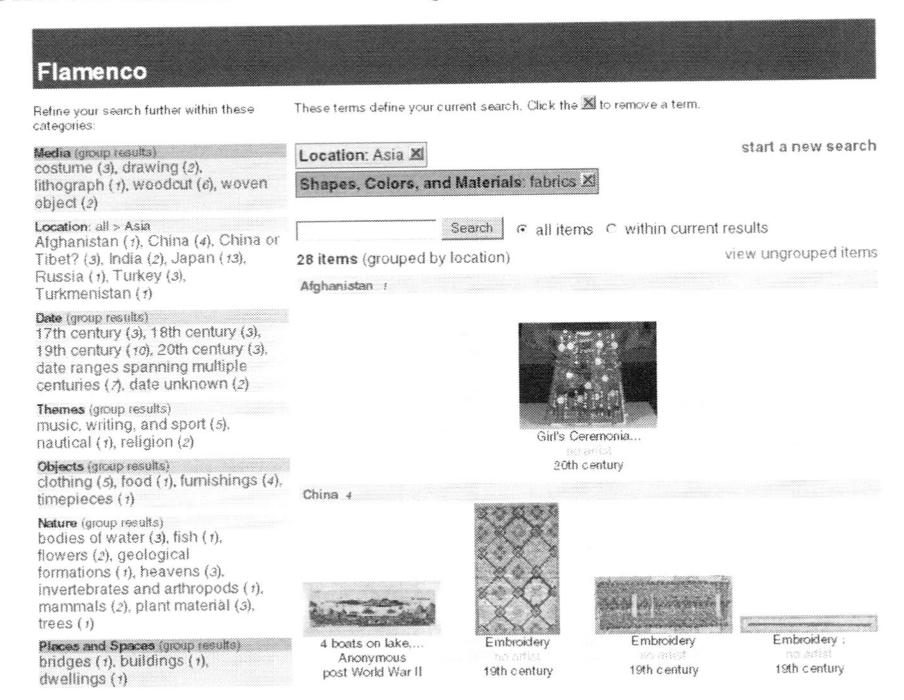

FIGURE 4.2: Flamenco interface with facets selected (Yee et al., 2003). Copyright © ACM 2003. Used with permission.

illustrates the selection of Baroque in the Era column, and mSpace has restricted the items that appear in the Composer column to those composers of the Baroque era.

In mSpace, slices are dynamic; they can be altered by rearranging, adding, or subtracting dimensions, enabling individuals to determine how to organize the domain in support of their interest. Terms in a given dimension (e.g., Baroque, symphony, serenade) may have little meaning to a novice in the subject area. mSpace provides "preview cues" (schraefel et al., 2003) to help address this problem by giving users a sense of each musical genre.

In addition to users exploring collections based on facets or clusters, it is also possible to expose more types of metadata at the interface. Searchers can partition collections based on criteria they find important (e.g., recency, edit history, content language). Following the application of these metadata filters, the extent of information space that requires manual exploration can be dramatically reduced, leading to massive efficiency improvements for exploratory searchers browsing the collection or interactively filtering and sorting the results (Belkin et al., 2001).

The Relation Browser (RB) is a general-purpose search interface that can be applied to a variety of data sets (Marchionini and Brunk, 2003). The RB aims to facilitate exploration of the

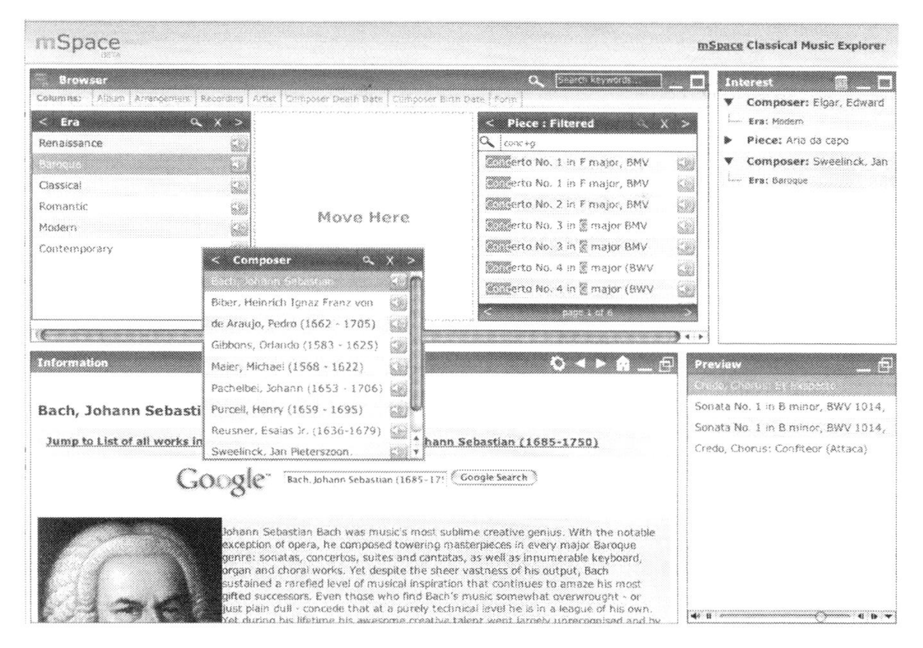

FIGURE 4.3: The mSpace Explorer (schraefel et al., 2006). A slice through the information space is shown with four dimensions: era, composer, form, and piece. The Form dimension is being dragged from the right of composer to the left, rearranging the slice. Illustration courtesy Max Wilson. Used with permission.

relationships between (among) different data facets, display alternative partitions of the database with mouse actions, and serve as an alternative to existing search and navigation tools. RB provides searchers with a small number of facets such as topic, time, space, or data format. Each of the facets is limited to a small number of attributes that will fit on the screen. Simple mouse-brushing capabilities allow users to explore relationships among the facets and attributes, and dynamically update results as brushing continues. Figure 4.4 shows a screenshot of the RB in action.

The current query is shown at the top of the display, and items in the current query are highlighted in red in other areas of the display. Facets and keyword searching allow users to easily move between searching and browsing strategies. Bars in the facet list indicate the number of results for the current query and the overall number of items in the collection that have this facet. Elements of the interface are coordinated and dynamic. This means that as users brush the mouse pointer over a facet, the elements update to show what the results would be after including the mouse-brushed item in the search. This feature allows users to quickly and easily explore the information space. Additional views are supported for both the results (display as a list or in a grid), and the facets (display in a list or as a "faceted cloud" similar to a tag cloud; Capra and Marchionini, 2008). RB depends on dynamic client-side graphics to be able to update the interface in real time. This creates issues of

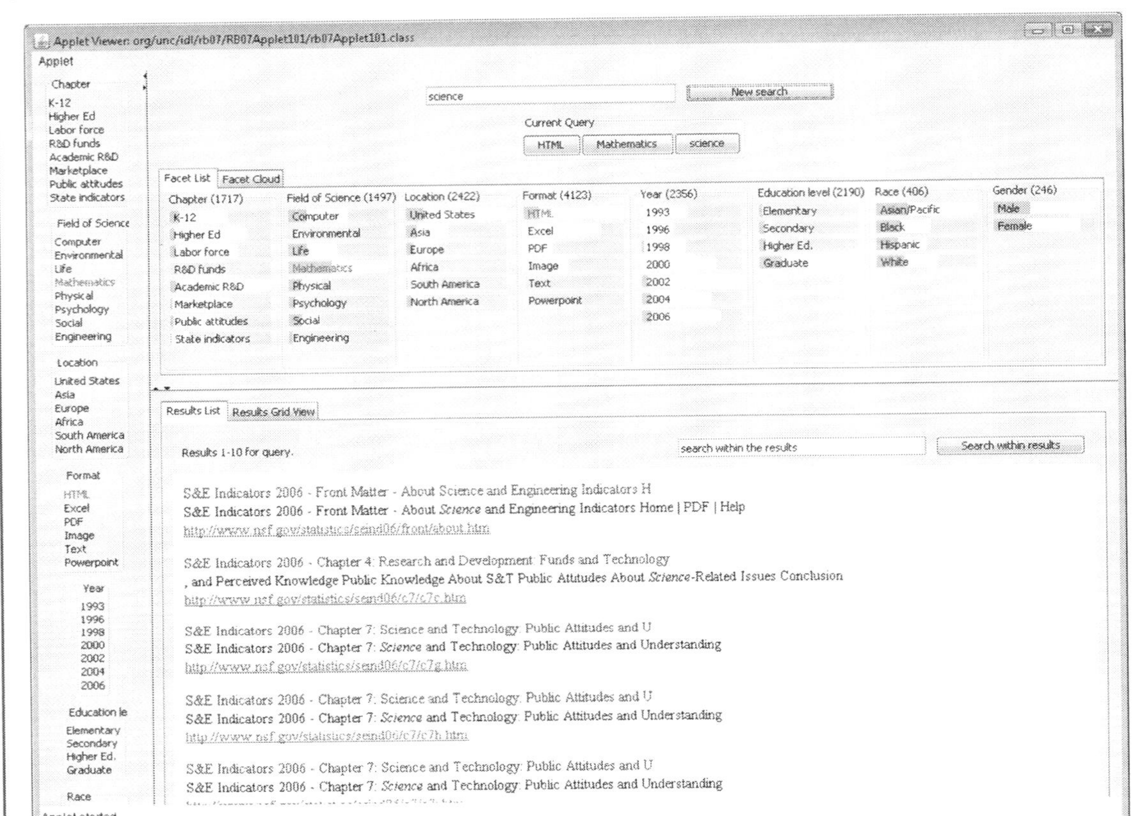

FIGURE 4.4: Relation Browser (Capra and Marchionini, 2008). Copyright © ACM 2008. Used with permission.

scalability that must be resolved before the RB and similar data-dependent tools can be deployed at Web scale, with potentially billions of records that must be manipulated instantly.

White and colleagues (2006) suggested that "an exploratory search may be characterized by the presence of some search technology and information objects that are inherently meaningful to users (for example, their images, email messages, and music files)." Personal information management (Jones, 2008), where people to acquire, organize, maintain, retrieve, and use information items, is a research area of growing importance. Although users may have encountered personal content previously, information overload may make finding and re-using that information similar to information discovery (Cutrell et al., 2006; Dumais et al., 2003; Ringel et al., 2003). However, similar concepts could also be applied to the prefiltering of search results in other domains, as is evident in query operators such as "filetype:" in popular search engines.

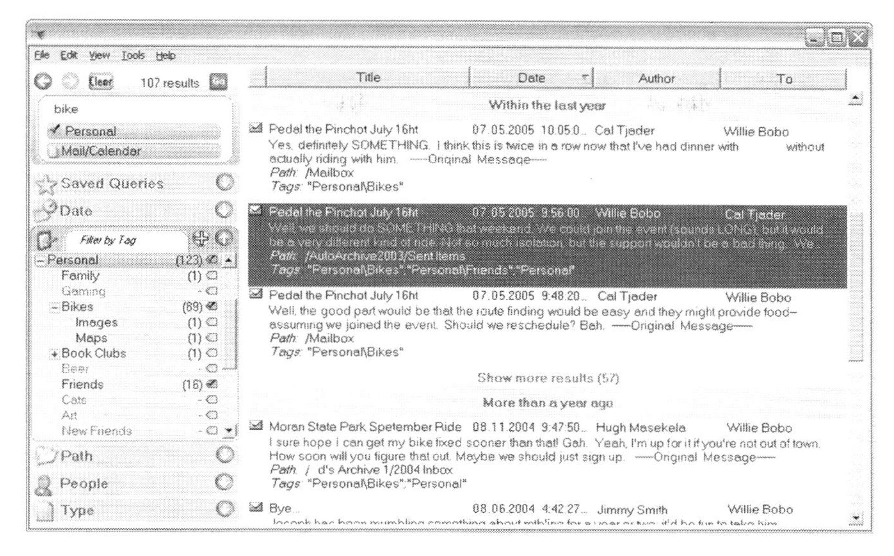

FIGURE 4.5: Screenshot of Phlat (Cutrell et al., 2006). Copyright © ACM 2006. Used with permission.

Similar to many successful faceted search interfaces, Phlat (Cutrell et al., 2006) combines keyword search and metadata browsing in a seamless manner, allowing people to quickly and flexibly find their own content based on desired result properties (Figure 4.5). In addition, Phlat provides a facility for tagging items with a uniform system of user-created metadata.

Key to the design of Phlat is the tight coupling of searching and browsing. To reinforce this unification, keyword and metadata search terms appear similar and are located in the same query box. From any broad starting point, a user may then rapidly filter, sort, and iterate on their query based on what they see and remember until they locate relevant information.

Stuff I've Seen (Dumais et al., 2003) is a system that facilitates search on personal content and information re-use by providing: (1) a unified index of information that a person has already viewed, whether it was viewed as an email, Web page, document, appointment, etc. and (2) search and interface technology that utilizes rich contextual cues such as time, author, thumbnails, and previews to search for and present information. In contrast, Web search results lack personal context, making rank the only reasonable alternative for ordering results.

As part of the Stuff I've Seen research, Ringel and colleagues (2003) demonstrate the value of a timeline visualization that capitalizes on the research in the psychology and human factors literature on landmarks and episodic memory (Czerwinski and Horvitz, 2002; Smith et al., 1978; Tulving, 1983). Results of searches are presented with an overview-plus-detail timeline visualization (Figure 4.6). A summary view shows the distribution of search hits over time, and a detailed view allows for inspection of individual search results.

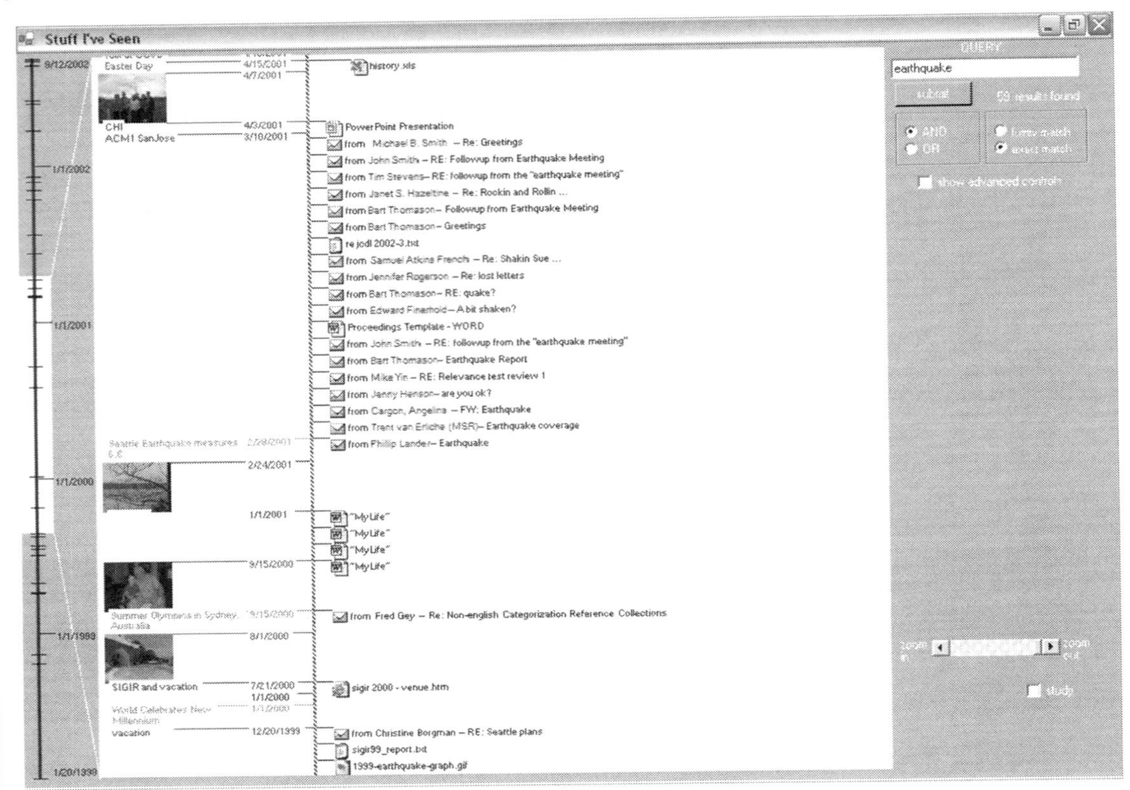

FIGURE 4.6: Screenshot of timeline visualization, based on Ringel et al. (2003). Illustration courtesy Merrie Morris. Used with permission.

The systems presented in this subsection highlight the potential of facets and metadata for supporting exploratory search activities in unseen collections or a user's personal content. It is likely that exploratory searchers would be able to name at least one attribute of relevant information, even if it is only language, filetype, or recency. This will increase search efficiency by restricting the extent of the collection that requires manual exploration.

4.3 LEVERAGE SEARCH CONTEXT

Tools to support result retrieval using contextual information are valuable because information needs during exploratory searches are ill-defined. Context can be used directly during search and retrieval for tasks such as: (1) query disambiguation (e.g., a query for "jaguar" may mean the car manufacturer or a species of animal; Glover et al., 1999); (2) query expansion based on analysis of the top-ranked documents (Xu and Croft, 2000); (3) result ranking in link analysis algorithms using anchor text; or (4) to support document selection through query-biased summarization (Tombros and Sanderson, 1998).

Context can be captured explicitly by asking searchers to mark useful queries or search results over time to build (e.g., Bharat, 2000) or to indicate useful text fragments (e.g., Finklestein et al., 2001), or by implicitly mining contextual information from users' interaction behavior (e.g., Dumais et al., 2004; Kelly and Belkin, 2004; Shen et al., 2005). The selection of a domain-specific search engine rather than a general-purpose search engine also provides valuable implicit contextual information (Lawrence, 2000).

Attentive systems such as Lira (Balabanovic and Shoham, 1995), WebWatcher (Armstrong et al., 1995), Suitor (Maglio et al., 2000), Watson (Budzik and Hammond, 2000), PowerScout (Lieberman et al., 2001), and Letizia (Lieberman, 1995) accompany the user during their information-seeking journey and model user interests by observing search behavior (and other behaviors in intermodal systems). Such systems typically operate on a restricted document domain or on the Web. The methods used to capture this interest and present system suggestions differ from system to system. Letizia (Lieberman, 1995) learns user's current interests and searches the Web automatically (i.e., predicting what searchers may be interested in the future, based on inference history) to recommend nearby pages. PowerScout (Lieberman et al., 2001) uses a model of user interests to construct a new complex query and search the Web for documents semantically related to the last relevant document. WebWatcher (Armstrong et al., 1995) also observes user browsing behavior, but goes beyond, acting as a learning apprentice (Mitchell et al., 1994). Over time, the system learns to acquire greater expertise for the parts of the Web that the user has visited in the past and for the topics of interest to previous visitors. Suitor (Maglio et al., 2000), tracks computer users through multiple channels such as gaze, Web browsing, and application focus to determine their interests. Watson (Budzik and Hammond, 2000) uses contextual information in the form of text in the active document and proactively retrieves documents from distributed information repositories via a new query.

The IR community created a medium of knowledge elicitation traditionally performed by human intermediaries. User models and task models can be created to be used in the selection of retrieval strategies (Belkin et al., 1993; Brajnik et al., 1987; Croft and Thompson, 1987; Oddy, 1977; Rich, 1983; Vickery and Brooks, 1987). Systems of this nature have focused on characterizing tasks, topic knowledge, and document preferences to predict searcher responses, goals, and search strategies. These systems typically make many assumptions about the search environment in which they operate and the searchers that use them. They can be useful in exploratory search scenarios given the depth of information that may be available about the user and the task, gathered over multiple search sessions.

In addition to developing models of user interests, it is valuable for systems to be aware of when users are experiencing difficulty during their searches (e.g., Horvitz et al., 1998). This is arguably an ideal time for system intervention with recommendations about queries to issue, pages to visit, or with suggestions about actions to assist the user in pursuit of their goals. It may also be useful in these circumstances to display explanations for why documents or actions were suggested based on the inferred task context. Fisheye or other peripheral views also provide users with clues about nearby information (e.g., Furnas, 1986).

4.4 OFFER VISUALIZATIONS TO SUPPORT INSIGHT/DECISION MAKING

Exploratory searchers should be allowed to select a task-appropriate form of data display (Shneiderman et al., 1997). Exploratory search systems must provide overviews of the searched collection and large-scale result sets to allow information to be visualized and manipulated in a variety of ways. Information visualization and the use of graphical techniques help people understand and analyze the data, and they are important during hypothesis generation. In contrast with scientific visualization, information visualization focuses on abstract data sets, such as unstructured text or points in high-dimensional space. It forms part of the direct interface between user and machine. Information visualization amplifies cognitive capabilities in six basic ways (Card et al., 1999): (1) by increasing cognitive resources, such as by using a visual resource to expand human working memory; (2) by reducing search, such as by representing a large amount of data in a small space; (3) by enhancing the recognition of patterns, such as when information is organized in space by its time relationships; (4) by supporting the easy perceptual inference of relationships that are otherwise more difficult to induce; (5) by perceptual monitoring of a large number of potential events; and (6) by providing a manipulable medium that, unlike static diagrams, enables the exploration of a space of parameter values.

Information visualization tools provide users with the ability to explore a range of data dimensions seamlessly. These capabilities of information visualization, combined with computational data analysis, can be applied to analytic reasoning to support the sense-making process and exploratory search. As is the case in exploratory data analysis, visualizations can be used to support the generation of hypotheses and decision making. Companies such as SAP (sap.com) and Spotfire (spotfire.com) have developed applications to process business intelligence data and help data analysts draw reasonable inferences. IBM's manyEyes (Viégas et al., 2007) shows the value of being able to share visualizations of data, by adding manipulable facets onto the visualization. manyEyes uses information visualization as a catalyst for discussion and collective insight about data.

Projects such as Lifelines2 (Wang et al., 2008) have used larger sets of data from patients' electronic health records and medical test results, enabling medical professionals to align-rank and sort them according to the attributes available on the data. Lifelines2 enables discovery and exploration of patterns across these records to support hypothesis generation, and find cause-and-effect

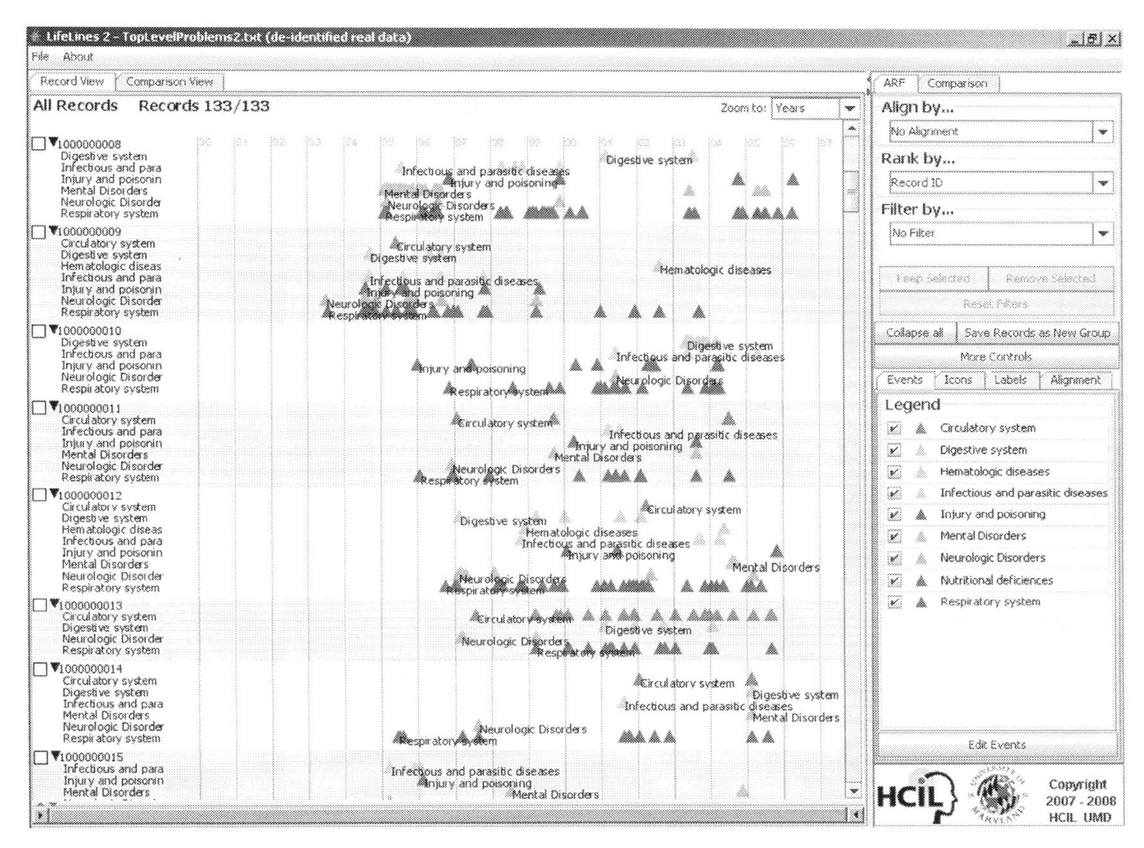

FIGURE 4.7: Interface to Lifelines2 with recorded incidences marked. Human–Computer Interaction Laboratory, University of Maryland, http://www.cs.umd.edu/hcil/lifelines2. Used with permission.

relationships in a population. Figure 4.7 shows Lifelines2 with recorded incidences of conditions marked. Figure 4.8 shows the align and zoom sequence.

Lifelines2 enables dynamic exploration of many "what if" scenarios and new discoveries through correlations to be made (schraefel, 2009). The need for such analysis is common in exploratory search scenarios (as searchers seek explanations and causative effects for observed phenomena), and as a result, exploratory search systems must allow the underlying data to be visualized and transformed in different ways.

4.5 SUPPORT LEARNING AND UNDERSTANDING

Systems in support of exploratory search activities have an obligation to help users learn more about the subject area in which they are searching and comprehend the information they encounter.

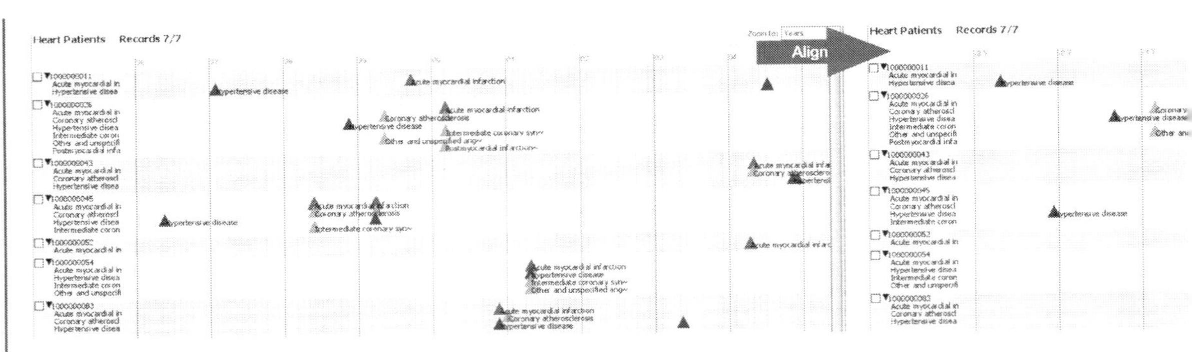

FIGURE 4.8: Performing align and zoom operations in Lifelines2. Human–Computer Interaction Laboratory, University of Maryland, http://www.cs.umd.edu/hcil/lifelines2. Used with permission.

Learning and understanding are important aspects of exploratory search that goes beyond IR; systems can no longer only deliver the relevant documents, but must also provide facilities for deriving meaning from those documents (Marchionini, 2006b). To augment intellect and help users make sense of encountered information, exploratory search systems target documents that contain topic overviews or appropriate content. The search experience is tailored to the individual searcher based on inferences made about result complexity and searcher-appropriateness given their estimated level of domain expertise (White et al., 2009). It is also possible to mine historical data sets of many users' search behavior and use the trails previously followed by expert users (and shared by them) as a way to educate novices about effective search strategies and useful resources.

Systems such as SuperBook (Egan et al., 1989) and SuperManual (Folz and Landauer, 2007) improve the usability of existing documents through computer-based enhancements. Such enhancements give users access to additional features that may be helpful in comprehending texts. Lessons can be drawn from the e-learning and intelligent tutoring communities (Corbett et al., 1997), and users purposely engaged in sustained reasoning activities during browsing. Related work in the hypertext community on the creation of guided tours (Trigg, 1988) is also useful to support learning and understanding.

Exploratory search systems that offer guided tours must place documents in ascending order of complexity or in an order most conducive to user learning, personalized to the user or to those similar in skill level. In response to informational queries, or at a branching point in the search, they would return guided tours and a result list. Systems must also take into account the authoritative nature of the document (e.g., how many citations has it received from other authors) and the reputation of the author if available. For example, authority is used in link-analysis algorithms used for Web search ranking (Kleinberg, 1999) and is one way in which the value of the document can be determined. Exploratory search systems must also offer topic coverage and controllable result

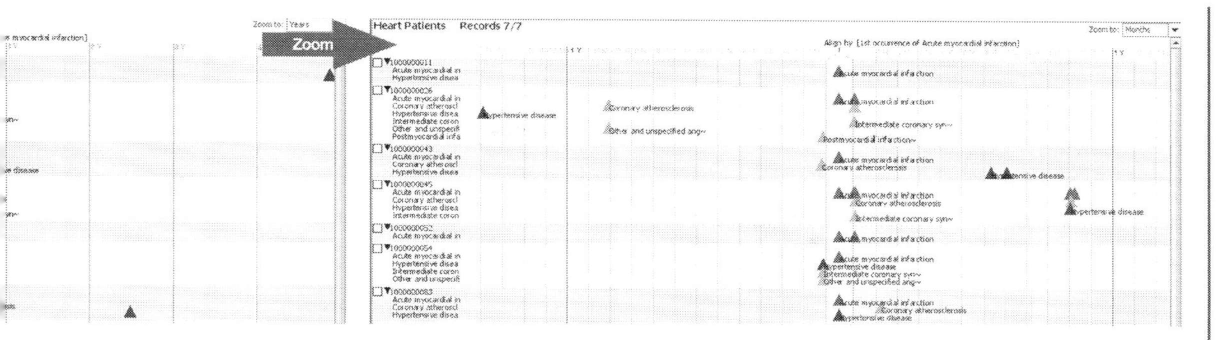

diversity to allow users to learn more about an entire subject area topic or focus their search on a particular subtopic.

4.6 FACILITATE COLLABORATION

Collaborative information seeking can be more effective than solitary information seeking: different people bring diverse perspectives, experiences, expertise, and vocabulary to the search process (Golovchinsky et al., 2009). A retrieval system that takes advantage of breadth of experience should improve the quality of results obtained by its users (Baeza-Yates and Pino, 1997). This is applicable during exploratory search scenarios when each user's understanding of the problem context may be vague, but when pooled, the salient aspects are identifiable. Pooling cognitive resources may also yield benefits in terms of coverage of the solution space; as more people bring with them ideas on complex problem solving.

There are many forms of collaboration in search, such as user interfaces that allow multiple people to compose queries (Morris and Horvitz, 2007) or examine search results (Smeaton et al., 2006), and community-based recommendation systems (Smyth et al., 2005). Several systems have explored interfaces that allow multiple users to collaboratively interact with online information, which is different from search itself. The Sociable Web (Golovchinsky, 1997), for example, allows a user to see that others are currently viewing the same web page and to communicate with those people. Alternatively, several systems allow users to share bookmarks or favorites lists, such as DogEar (Millen et al., 2005), WebTagger (Keller et al., 1997), Wittenburg and colleagues' system (1995), and the commercial site del.icio.us.[2]

Morris and Horvitz (2007) introduce SearchTogether, a prototype that enables groups of remote users to synchronously or asynchronously collaborate when searching the Web. This type of collaboration may be most effective in exploratory searches. Not only can the task be divided among all collaborators, but collaborators can learn from each other in real-time and pose questions

[2] http://del.icio.us.

to other searchers if necessary. Figure 4.9 shows a screenshot of SearchTogether with the main components highlighted and described in the caption.

Pickens and colleagues (2008) explored the possibilities of synchronous, explicit, algorithmically mediated collaboration for search tasks. They describe a retrieval system wherein searchers collaborate explicitly (intentionally) with each other in small, focused search teams, rather than implicitly with anonymous crowds. Collaboration goes beyond the user interface: information that one team member finds is not only presented to other members in pursuit of shared learning, but used by the underlying system in real-time to improve the effectiveness of all team members, while allowing each to work at their own pace.

Exploratory search systems need to utilize collaboration between searchers attempting the same task, either at the same time or with latency from delays between Web page postings or bookmarking activities. Individuals may know each other before the task begins or be matched by the system once their interests become clear. Searching as part of a group with common goals and interests is a mutually beneficial activity that can help all members navigate a complex information landscape more effectively.

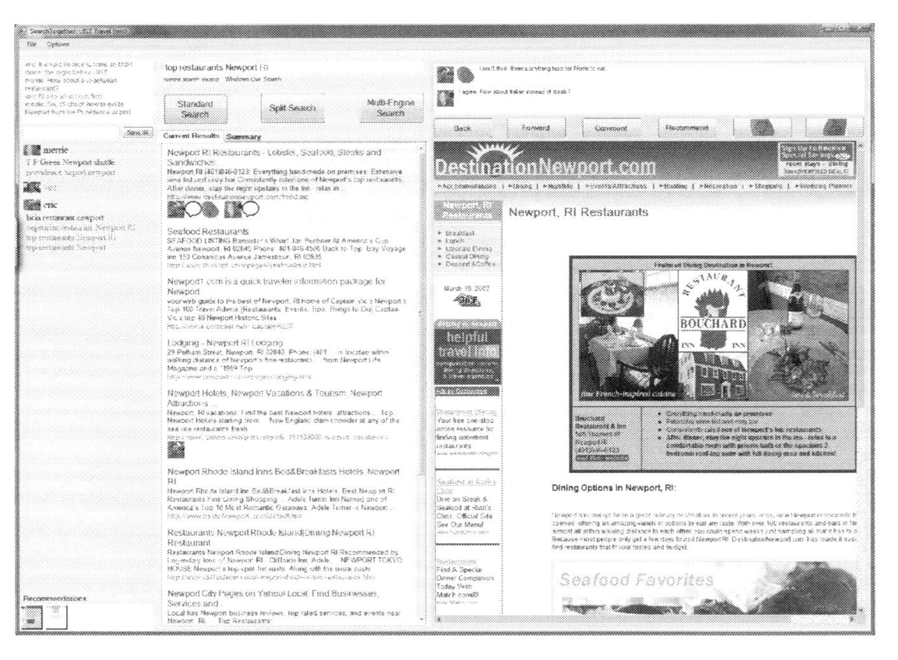

FIGURE 4.9: Screenshot of SearchTogether, based on Ringel and Horvitz (2007). Illustration courtesy of Merrie Morris. Used with permission.

4.7 OFFER HISTORIES, WORKSPACES, AND PROGRESS UPDATES

Exploratory searches typically involve the examination of multiple information sources and transcend multiple search sessions. Exploratory searchers require tools that allow them to easily revisit previously encountered items and to store potentially useful information for later use. Knowledge of how much of the information space has been explored on a topic and what remains to be seen, is useful for exploratory searchers. Exploratory search systems should: (1) offer a smart and structured history, records of paths users followed to get to findings, and easy revisitation of results; (2) contain "workspaces" to support a spectrum of activities, from unstructured note-taking to integrated authoring environments; and (3) keep track of user progress (and alert them if they are straying), remember dead ends, and record what has already been seen.

Systems have already been developed that offer elements of histories, workspaces, and progress updates. Hunter Gatherer (schraefel et al., 2002) is an interface that lets Web users carry out three main tasks: (1) collect components from within Web pages; (2) represent those components in a collection; (3) edit those component collections. InkSeine (Hinckley et al., 2007) is a tablet computer search application that enables users to store a pointer to a search via a breadcrumb object intermixed with their handwritten notes. Dontcheva and colleagues (2006) developed a system for summarizing personal Web browsing sessions that allows users to define patterns for extracting structured information from a set of Web pages. SearchPad (Bharat, 2000) allows a user to explicitly flag a Web page for inclusion in a workspace in order to help the user maintain context during complex search tasks. Google's Notebook application[3] allows users to collect snippets of content from several Web pages and combine them in a single document.

White and colleagues (2006c) proposed the use of searcher-constructed concept maps to support oral history search. Oral history archives are rich in named entities and inter-entity relationships that can be tagged and made accessible to a search system. Concepts maps containing these entities and relationships may therefore be a reasonable way to facilitate search and use in these archives. Figure 4.10 shows an example of a concept map. It has been constructed interactively and maintains its state for an entire search session, or longer, if explicitly saved by the searcher. Searchers can annotate nodes in the map and create relationships between them.

Concept maps allow searchers to build a representation of their interests that may be helpful to their search. Searchers can store information fragments on a canvas, link these fragments interactively to create a concept map, and use the map to drive future searches or help to better understand their information problem.

[3] http://www.google.com/notebook.

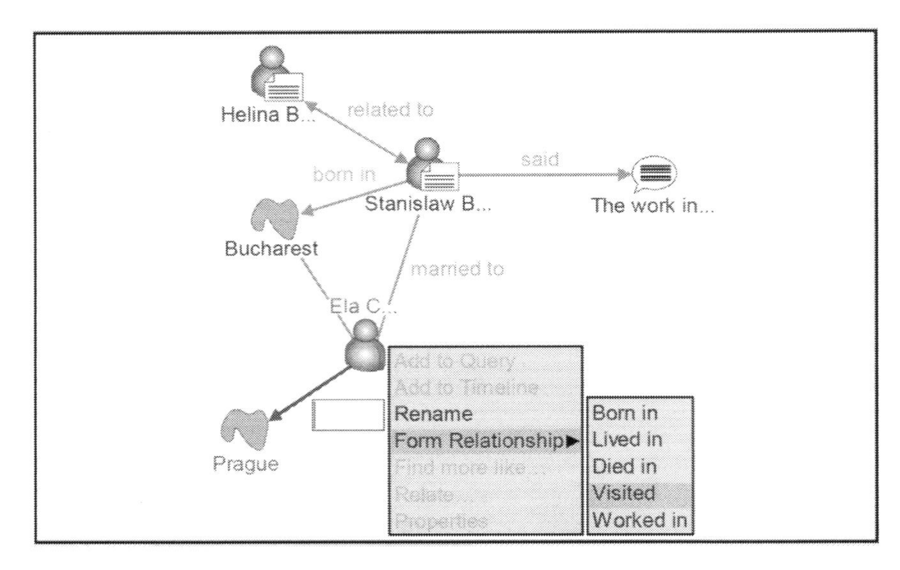

FIGURE 4.10: Concept map interface from White et al. (2006c). Copyright © ACM 2006. Used with permission.

Not all information encountered during exploratory searches is relevant. Exploratory search systems must allow users to easily gather information during the search and collate that information into cohesive summaries during or following the search. The information-gathering process and summaries created will drive insight generation, support decision making, and facilitate user learning. Through monitoring result features, such as the proportion of novel information in result sets relative to the information encountered, systems can alert users when they are exhausting an information resource or a line of inquiry and suggest a new information patch within which to forage or a new search direction.

4.8 SUPPORT TASK MANAGEMENT

Since exploratory searches likely transcend multiple search sessions, it is important that exploratory search systems provide a mechanism for users to save their state and allow them to return to previous search sessions later. The state not only includes the documents viewed, but includes all other contextual variables such as queries issued, relevant documents marked, paths followed, and potentially other applications opened. In exploratory search systems, tasks are important in the same way that queries are important in today's search systems. Exploratory searchers should be able to retrieve tasks to help find solutions to previously encountered problems, use task interactions as the basis for future retrieval, and share tasks and task outcomes with other users. There has been some progress in this area, and SearchBar (Morris et al., 2008) is an example. SearchBar is a system for proactive and persistent storage of query histories, browsing histories, and users' notes and ratings in an interrelated fashion. It supports multisession investigations by assisting with task context resumption and rediscovery of information.

4.9 SUMMARY

In this chapter, we have presented a number of features of exploratory search systems. The unique nature of searches where the target is unknown calls for new system designs to help users. Systems that support user exploration are a vital and a missing part of most current mainstream search technology. The systems described in this section offer a range of interaction modalities. However, seldom does one system offer more than a single mode of interaction. When conducting an exploratory search, it may be necessary for searchers to employ multiple interaction modes such as textual queries, query-by-example, facets/selections, dynamic queries, and guided tours to obtain the new understanding they seek. These techniques should exist harmoniously at the interface, with a balance between analytic and browsing strategies. To support intelligence amplification, exploratory search systems need to increase user responsibility as well as control; they must require human intellectual effort and must reward users for effort expended.

In addition to the eight features described in this chapter, exploratory search systems should be engaging and fun to use, as well as support modification by end users and information professionals. In addition, they should have flexible architectures to support rapid refinement and expansion, and be integrated into the information ecology rather than acting as discrete stand-alone services (Marchionini, 2006b).

• • • •

CHAPTER 5

Evaluation of Exploratory Search Systems

When evaluating exploratory search systems (ESSs), it is impossible to completely separate human behavior from system effects because the tools are so closely related to human acts, they become symbiotic. This symbiosis is intentional; exploratory search systems act as cognitive prosthetics, and must be closely coupled to the user and their intentions. Information seeking is usually intertwined with many activities, and it is common for users to be engaged in multiple information-seeking tasks simultaneously (Kelly et al., 2009).

Recently, researchers have focused on developing new systems and interfaces to support exploratory search activities, rather than their evaluation. It is also necessary to understand the behaviors and preferences of users engaged in exploratory searching, tasks supported by ESSs, and measures of exploration success. While search systems are expanding beyond the support of simple lookup into complex information-seeking behaviors, evaluation of search systems has remained limited to those that encourage minimal human–machine interaction.

IR is by nature an experimental discipline; the evaluation of retrieval algorithms and other aspects of system design such as document indexing and the user interface are central to progress in the field. The Cranfield methodology (Cleverdon et al., 1966), which was later used by the NIST-sponsored Text Retrieval Conference (TREC) (Harman, 1993; Harman and Voorhees, 2005), has been a useful paradigm for the objective comparison of IR systems, where only one aspect of a system is varied at any point in time. TREC is an ongoing series of workshops on a range of different IR research areas. Its objective is to support and encourage research within the IR community (mainly on the development of ranking algorithms) by providing the infrastructure necessary for large-scale evaluation of text and multimedia (e.g., image and video) retrieval methodologies.

TREC provides a medium for the evaluation of algorithms underlying the analytical aspects of IR systems, yet it struggles because the experimental methods of batch retrieval are not suited to studies of how search systems are used by human searchers. Search systems are not used in isolation from their surrounding context, they are used by people who are influenced by environmental and situational constraints such as their current task. Effective search systems must have provisions to adapt to these contextual constraints (Ingwersen and Järvelin, 2005), and evaluation methodologies must be capable of analyzing systems on the basis of constraints. Since TREC-3, the conference has extended its mandate to recognize the importance of the user in information-seeking.

The TREC Interactive Track (summarized by Dumais and Belkin, 2005) and later the TREC High Accuracy Retrieval of Documents (HARD) Track (Allan, 2003), have both attempted to bring the user into the evaluation process. However, these tracks struggled to establish comparability between experimental sites, in terms of the experimental systems devised and the measures used. They were also adversely affected by the dependence on relevance judgments and interactions between users, tasks, and systems. Nonetheless, the Interactive Track was successful at highlighting the importance of users in information-seeking (Lagergren and Over, 1996). It is unclear if the evaluation of exploratory search will blossom within the TREC paradigm; however, researchers are increasingly turning their attention toward new ways to systematically investigate ESS effectiveness and the information-seeking process.

High levels of interaction, integral to exploratory search, pose an evaluation challenge: there is potential for confounding effects from the different exploration tools, the desired learning effect is difficult to measure, and the potential effect of fatigue limits evaluation to a small number of topics. All of these attributes make it difficult to achieve the statistical significance required by a meaningful quantitative analysis. A key component of exploration is human learning, a topic studied extensively by cognitive psychologists (e.g., Landauer, 2002). Subject-matter learning is a viable way to evaluate ESSs, as a function of exploration time and effort expended.

Support for more-rapid learning across a number of users and a range of tasks is indicative of a system that is more effective at supporting exploratory search activities. For example, in evaluation of Scatter/Gather, an interface designed to support search result exploration through text clustering, Pirolli and colleagues (1996) measured user learning and understanding in terms of topic structure and query formulation capabilities at various points during subject interaction with the system. In comparison to a control group that performed the same tasks using a standard search engine, users of the Scatter/Gather system showed larger gains in understanding the underlying topic structure and in formulating effective queries. Similarities between exploratory search, sense-making, and information foraging signify that an analysis of the costs involved in the process in terms of gain for time spent representing/understanding the task and finding/selecting information may also be useful for comparing exploratory search systems (Russell et al., 1993). Ultimately, researchers studying exploratory search must measure the depth and effectiveness of learning rather than focus on efficiency. Time may be less appropriate as a measure of outcome. However, learning time may be a reasonable metric of choice for exploratory search at this stage in the development of ESSs.

The evaluation of ESS is not substantially different from the evaluation of other highly interactive systems. Subjective measures such as user satisfaction, engagement, information novelty, and task outcomes are important, but it is through measurement of interaction behaviors, cognitive load, and learning that one can truly evaluate the effectiveness of ESSs. The approach adopted at TREC has led to the rapid development of effective ranking algorithms for document retrieval. As a result of such research, search systems such as Google, Yahoo!, and Live Search cope well with navigational requests (e.g., find a specific person's homepage) and closed informational requests (e.g., answer to a question which has a single answer). For more challenging information-seeking tasks, recall is as important as precision, and it is critical that evaluation of information-seeking support systems uses recall (Tunkelang, 2009). It has been suggested that repositories of data and tasks (similar to TREC) could be used to evaluate ESS based on information visualization (Plaisant, 2004).

5.1 METRICS

Evaluation metrics facilitate the incremental improvement of search technologies by providing a way to assess system performance and facilitate comparisons between experimental systems. Devising such metrics in exploratory search is particularly challenging, since the goals are subject to change as the searcher interacts with the system. In such cases, standard precision-recall metrics may be ineffective (although there has been some consideration given to devising variants of these metrics more suitable for IIR (Borlund and Ingwersen, 1998; O'Brien, 2008).

In Chapter 1, we quoted Douglas Engelbart (1962), suggesting that the increased capability resulting from augmenting human intellect would likely lead to: "more-rapid comprehension, better comprehension, the possibility of gaining a useful degree of comprehension in a situation that previously was too complex, speedier solutions, better solutions, and the possibility of finding solutions to problems that before seemed insoluble." To evaluate exploratory search systems, we must target the longer-term effect on the user of using the cognitive prosthetic as well as their current task performance. Process-specific measures of learning, cognitive transformation, confidence, engagement, and affect are important, as well as result relevance and utility across multiple query iterations and search sessions.

A workshop entitled "Evaluating Exploratory Search Systems," organized by Ryen White, Gary Marchionini, and Gheorghe Muresan, held in conjunction with the 2006 ACM SIGIR Conference on Research and Development in Information Retrieval, brought together experts from academia and industry to discuss exploratory search evaluation (White et al., 2006b). The following are some candidate metrics that emerged from brainstorming and breakout sessions at the workshop which may be used to assess the performance of exploratory search systems:

1. Engagement and enjoyment: The degree to which users are engaged and are experiencing positive emotions can be a strong indicator of system performance. Amount of interaction required during exploration, extent to which the user is focused on the task, and content with the system's response can indicate whether the system is fulfilling its role in supporting search activity. The number of actionable events (purchases or forms filled, bookmarking, feedback or forwarding events, etc.) can be used as a metric to approximate levels of engagement and enjoyment.
2. Information novelty: Since the goal of exploration is to encounter information not seen before, it is appropriate to include the amount of new information encountered as a way of measuring effectiveness of an exploratory search system. The rate at which users encounter new information is an important aspect of determining how effectively exploratory search systems provide users with new information.
3. Task success: Task success should not only be based on whether the user reaches a particular target document, but also on whether they were able to encounter a sufficient amount of information and detail en route to reaching their goal. As one workshop participant remarked "[exploratory search] is more about the journey than the destination." Since task success may be based on the difficulty of the task, metrics such as the clarity measure (Cronen-Townsend et al., 2002) may also be appropriate.
4. Task time: Time spent to reach a state of task completeness is an effective way to assess efficiency of exploration activities. Task time can include total time spent, time spent looking at irrelevant documents, and proportion of time spent engaged in directed search versus amount of time spent exploring. Task completeness would be indicated by experimental subjects based on their own perceptions of their task state.

    To target learning, experimenters can monitor how much subject-matter learning is achieved as a function of exploration time. Table Lens (Rao and Card, 1994) is an information visualization tool that support sense-making from large tables or spreadsheets. Pirolli and Rao (1996) studied the rate at which participants grasped properties of variables and relationships among variables in Table Lens compared to the table representation; they showed that Table Lens helped users learn at a faster rate. Pirolli (2007) suggested that exploratory search systems could be evaluated through cost structure analysis by finding metrics of learning or expertise and then by comparing how exploration with one system versus another produces better or worse gains against those metrics (see Figure 5.1). It may not be possible to compute a goal for each task. In such cases, one must compare searcher

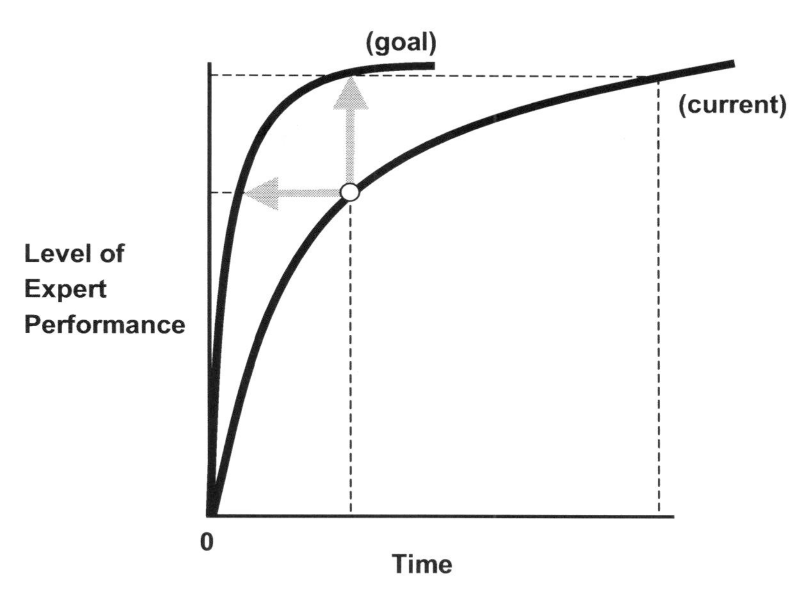

FIGURE 5.1: Measuring performance relative to optimal (based on Pirolli, 2007).

knowledge before and after the task, ask them for feedback about their experience, and focus on their perceptions of task completeness.

Measuring how quickly users reach a particular state of knowledge may contradict other measures of system performance such as engagement, enjoyment, and learning. In such cases, it may be in the interests of users to maximize rather than minimize time spent on task.

5. Learning and cognition: Learning is key to exploratory search. Through measuring cognitive and mental loads, the attainment of learning outcomes, the richness/completeness of a user's postexploration perspective, the amount of the topic space covered, and the number of insights users acquire, we can compare exploratory search systems in terms of learning and cognition. In addition to studying individual learning, shared learning can also be studied through the state or state change in social networks. For example, an increase in the traffic to key items on a subject within a group of Facebook[1] or Friendster[2] users may be indicative of shared learning in that group.

[1] http://www.facebook.com.

[2] http://www.friendster.com.

Each of the objective and subjective metrics identified in this section tackle an important aspect in the evaluation of efficiency and effectiveness of exploratory search systems, as well as users' perceptions of their usefulness. However, in this section, the metrics have been presented independently. In practice, exploratory search systems affect multiple aspects of information-seeking behavior, and it is important to note that a variety of metrics will need to be employed in tandem to evaluate such systems. The "informativeness" measure (Tague-Sutcliffe, 1992) took a first step in this direction by combining subjective user responses regarding information utility with a penalty derived from the system's ability to return relevant items, ranked in descending order of relevance. Toms and colleagues (2005) proposed the use of factor analysis (FA), and O'Brien (2008) proposed the use of structural equation modeling (SEM) as a way to examine the interrelationships between multiple search system evaluation metrics, allowing different types of data (e.g., user attitudes, observed behaviors, system performance) to be examined simultaneously for a more holistic approach to evaluating exploratory search system performance.

Given the complexity of exploratory search systems in terms of interface features and interaction supported, it is important to consider the relationship between varied and multiple measures during their assessment. Techniques such as FA and SEM will assist with evaluation, but the challenge remains in how to interpret the output of such statistical techniques. Challenges include the heuristic approach that must be employed for FA and the difficulties in developing models that allow causal inference to be established. The movement from the findings of such techniques to actionable design implications or an understanding of how exploratory search systems are being used and their comparative performance has yet to be established.

5.2 METHODOLOGIES

The role of evaluation in exploratory search is primarily to assess the success of the information-seeking process at reaching the information target(s) for the current session, if those exist, and achieving higher-order learning objectives for the searcher, such as the ability to apply their gained knowledge to related situations or to design a new product, resulting from knowledge synthesis. Evaluation methodologies are tightly connected to how user interaction behavior is represented and to the metrics adopted for measuring success. Interaction models specify the most representative or relevant factors of interactions in a certain context. Metrics represent both a conceptualization of the models and a measure of retrieval success. As such, evaluation methodologies connect models and metrics by specifying the rules, methods, and assumptions employed in evaluation, as well as rationale and philosophy behind the evaluation. It is natural for an emerging research community to be tentative in developing or adopting an evaluation methodology, as is the case with research in exploratory search systems. In exploratory search, researchers have adopted experimental settings, document collections, and investigation methods from areas such as IR, human–computer interaction, and psychology.

Effective evaluation of exploratory search systems requires researchers to first learn about the range of information-seeking tasks, processes, and search strategies which users engage in during exploratory search scenarios. The findings of these investigations can help with the design of laboratory evaluations, the representation of previously ignored aspects of information seeking (such as the information environment), and the identification of important research questions and system design needs. In exploratory searches, process of searching is just as important—if not more important—than the location of the information target. Finally, as stated earlier, exploratory search takes place over sustained periods of time, and this implies that longitudinal evaluation designs which measure change are most appropriate (Kelly et al., 2009).

In the evaluation of interactive search systems, data can only be collected from small numbers of users and about small numbers of tasks. Small sample size limits the generalizability of such studies' findings. The use of complimentary methods, such as laboratory studies, log analyses, and ethnographic observations provide clarity in understanding how systems support users in the search process (Grimes et al., 2007). Without the use of these methods, exploratory search system evaluation will require the development of longitudinal research designs involving larger numbers of more diverse users. One way to address the need for such research designs is to create a "living laboratory" on the Web that contains evaluation resources and infrastructure for bringing researchers and users together (Kelly et al., 2009).

Crowdsourcing marketplaces, such as Amazon's Mechanical Turk,[3] are emerging as a popular way to obtain relevance assessments for IR experimentation (Alonso et al., 2008). Crowdsourcing can also be used to solicit research participants for exploratory search evaluation. To do so, the community will need to develop economic models to incentivize participation and develop infrastructure to recruit, retain, and experiment with participants. Parallel "flighting" of different user experiences is already common practice in many of the large online retailers and search engines companies. These companies already have developed or are developing experimental platforms that allow controlled large-scale experimentation. Each parallel flight is given a small fraction of user traffic to the site (usually around 1% initially and increasing depending on effectiveness), and metrics such as click-through rate (search and online advertising) and revenue/conversions (online advertising and retail) are computed.

More intensive collaborations between academia and industry may provide a vehicle to share experimental exploratory search systems with the masses and compute usage statistics on their performance. However, intellectual property and interface quality issues will need to be resolved before this collaboration can happen. Nonetheless, it is likely that, through their interaction behaviors, the user population will soon be directing whether experimental exploratory search systems built by partnerships between academia and industry are successful.

[3] http://www.mturk.com.

The goal of reaching methodological rigor in studying exploratory search has not yet been reached. However, as the field matures, a set of methods accepted by the research community is expected to emerge. In order to reach these goals, a wide variety of candidate methods must be employed, with the expectation that the best methods and practices will eventually prevail. For example, naturalistic, longitudinal studies should be employed alongside lab experiments in controlled conditions. Both of these techniques are useful for different reasons. Naturalistic, longitudinal studies are better suited to observe the information seeker's behavior and search strategies, as well as changes in information needs and behavior that occur over time. Moreover, they are invaluable in developing and testing interaction models, and in ensuring that assumptions in user models hold, in general, or in certain contexts. In contrast, qualitative laboratory studies have the advantage of comparability and repeatability, and they support quantitative studies that attempt to answer research questions about the level of support that different system components offer to the user. Combinations of qualitative and quantitative evaluation methods have already been used for system evaluation (e.g., Toms et al., 2003; Yee et al., 2003).

Building appropriate test collections remain a difficult challenge: most existing experimental settings are based on the assigned task paradigm and on the assumption that the information need is static during the interaction. Research on task development (such as that of Borlund (2003) and more recently by Kules and Capra (2008) allows for the creation of simulated work task situations that are well-suited for exploratory search situations, as they are comparable between experimental subjects, but allow for personal assessments of relevance. Designing tasks to study exploratory search can be difficult because of the need to induce an exploratory rather than directed style of search. Also difficult is the need for tasks to be constructed in such a way that the results can be compared between subjects in a single study and across multiple studies by different research groups.

Kules and Capra (2008) identify a set of desirable characteristics for exploratory search tasks and propose a formal procedure for constructing tasks. The procedure draws task topics from query log data, integrates them into a high-level work scenario (Borlund, 2003), and addresses practical issues encountered in controlled or semicontrolled evaluations. Experimental evaluation of four tasks created using this task-generation procedure suggest that the procedure led to well-grounded, realistic tasks which elicited exploratory search behavior. The characteristics propose that an exploratory task: (1) indicates uncertainty, ambiguity in information need, and/or need for discovery; (2) suggests knowledge acquisition, comparison, or discovery task; (3) provides a low level of specificity about the information necessary and how to find the required information; and (4) provides sufficient imaginative context in order for the test persons to be able to relate and apply the situation.

Aside from improving the development of search tasks, exploratory search evaluation methodologies should take into account learning that takes place during the search session, evolution of the information need, dynamic nature of relevance judgments, as well as personality, background, knowledge, and preferences of the searcher. The addition of these features to the current Cranfield model of IR evaluation may lead to large increases in costs or reductions in power for IR experimentation (Voorhees, 2007). However, it has also been shown that improved relevance ranking may not always translate into better task performance (e.g., Hersh et al., 2000). A compromise between the many demands on any alternative evaluation methodology and the Cranfield model may be the simulation of searcher interaction behavior (e.g., White et al., 2005; Lin and Smucker, 2008). Simulations can be developed on interaction log data or based on cognitive models such as ACT-R (Anderson et al., 2004) and can incorporate features of user interaction behavior. The advantages of simulations include their repeatability and coverage of all interaction permutations.

5.3 SUMMARY

In this chapter, we have discussed opportunities for the evaluation of exploratory search systems. Traditional measures of IR performance based on retrieval accuracy may be inappropriate for the evaluation of these systems. The use of metrics based on engagement, enjoyment, novelty, task time and success, and learning provides an opportunity for understanding exploratory search system performance and for the comparison of different systems. Exploratory search evaluation methodologies must include a mixture of naturalistic and longitudinal studies. Highly refined user simulations developed based on interaction logs or cognitive modeling may serve as a compromise between existing IR evaluation paradigms and the enhancements required to support exploratory search evaluation.

• • • •

CHAPTER 6

Future Directions and Concluding Remarks

6.1 FUTURE DIRECTIONS

User expectations of search systems are exceeding systems' current capabilities. There is a need for search systems to provide electronic and dynamic mechanisms to help users tackle their search tasks more effectively and support long-term personal development goals such as intelligence amplification. As we have demonstrated in this lecture, exploratory search is emerging as an important discipline. Search system users can expect to witness a range of developments in the design of systems to support exploratory search activities. In this chapter, we present a few of the predicted advances in search technology.

6.1.1 Novel Interaction Paradigms

Exploratory search systems will utilize significant technological advances to support human–machine symbiosis during the search process. Users will not be restricted to using a desktop computer and a mouse pointer to manipulate information displays. Touch-sensitive tabletops or displays (e.g., Wilson, 2005; Wigdor et al., 2007), immersive environments (e.g., Cruz-Neira et al., 1992), holographic projections, eye tracking, speech recognition, sensors, and mobile devices will help users interact more fluidly with search technology and explore the information spaces of the future. Figure 6.1 shows an example of a large, multidimensional gestural display based on the 2002 Steven Spielberg movie *Minority Report.*

In the example in Figure 6.1, users can manipulate applications through fluid hand gestures, communicate with the system via verbal protocols, and visualize/organize available information. These features allow users to control many applications simultaneously and to rapidly survey the information landscape. As a result, they become more cognitively invested in the search process, can address mutiple aspects of the search task in parallel, and gain better perspective of their situation

FIGURE 6.1: Immersive, multitouch, gestural information display.

relative to the environment. These attributes are essential for rapid task completion and intelligence amplification.

6.1.2 Context Awareness

In a similar way to many other search activities, exploratory search activities occur within a work task context. At present, search is typically regarded as a means to find the necessary information to complete an aspect or aspects of a task. Given the dynamic and uncertain nature of exploratory searches, it will be necessary to embed support for exploratory search in many existing desktop applications. Search is becoming a first-order activity and has been integrated directly into the Microsoft Windows Vista and Mac OS X operating systems, as well as many applications such as the Microsoft Internet Explorer Web browser and Microsoft Office suite. Coupling information search with use allows systems to take advantage of knowledge about users' immediate task context (and less immediate context communicated through a semipermanent user profile) to tailor search results or user experience. Over time, the search system could also keep track of the users' current search skill and knowledge level and adapt the search results displayed to its estimate of the user's current level.

6.1.3 Task Adaptation

There are variety of different task types, and those that require exploration are only a subset. However, it is likely that all searchers will engage in at least one exploratory search per week, and for many, this activity will occur more often. For this reason, systems offer users the ability to explicitly request support for exploratory search should the task demand it. General Web search engine interfaces will offer search results, as they do today for navigational queries. These engines will also provide support for exploratory search tasks through different user interfaces and meet the requirements outlined in Chapter 5. General-purpose search systems will no longer suffice for the complex search tasks in which users engage. These systems may rely on users to select the most appropriate interface for their current task, or the search interface could make recommendations.

6.1.4 Decision-Making Support and What-If Analysis

Exploratory search systems must provide users with the ability to reason about the data they view to support decision making. Systems that target decision making will offer overviews and summaries of the data, dynamic queries, and "what-if" analyses. These systems will allow users to see the possible effects of their decisions and assign probabilities to each of the outcomes. In the context of search, decision-making support tools will help users select the optimal paths to follow through the information space. Systems will also gather information from disparate sources to provide users with enough information to make decisions regarding the task at hand. One of the most important decisions that users make when engaged in exploratory search activities is completion of the search task. Exploratory search systems will support choices about the finality of one's search by offering details on subtopics yet to be explored. Subtopics will be identified through automatic clustering of documents viewed and related documents based upon a crawl of corpora such as the Web.

6.1.5 Beyond the Personal Computer

Technological advances in large-screen displays, tactile user interfaces, virtual reality, and mobile devices are creating a wealth of opportunity for exploratory search systems to expand beyond the desktop computer and into our environment (simulated or otherwise). The ability to manipulate large volumes of data with a subtle hand gesture or to virtually immerse oneself in the corpus, opens up a new range of possibilities for our ability to interact with information and for the tools to search this information. Exploratory search via the personal computer will become one way in which exploratory search will be realized by system designers. Exploratory search support can already be found in devices such as the television (to explore and select movies to view on demand) and hi-fis (to select radio stations or songs to download from remote servers). Support for exploratory search will become more pervasive. For example, a home refrigerator could have a digital panel that allows people to explore recipe ideas based on current refrigerator contents as identified by radio-frequency identification tags or barcode scanning. On the move, personal music players will help people find and purchase songs from online collections, and personal digital assistants will download tours of previously unvisited cities or tourist attractions from the Web, showing them to people based on their global positioning coordinates and time/transportation constraints.

6.1.6 Collaborative and Social Search

Although search is often a solitary activity, the search task often involves multiple individuals. As such, it may be in the searcher's interests to collaboratively explore the information space and participate in shared learning. Aspects of the task can be allocated to different individuals or groups, making task completion more efficient. However, division of the task has the potential to hinder aspects of the learning for team members, making the attainment of a shared learning objective difficult. Collaborative exploratory search systems will provide a way to summarize (or facilitate rapid access to) already encountered information. The systems could tailor these summaries (or sets of links) to the respective skill levels of team members. This would allow the team to move rapidly toward task completion with minimal interruption from backtracking in review of information encountered by other team members. Immersive chat-rooms with high-quality streaming video and audio will let users converse in real time with those with similar interests and goals, from remote locations.

6.1.7 Learning About New Domains

Aside from learning about a new topic, the exploratory search systems of the future will help train users to search more effectively within a new domain. This may involve using explicit annotations or the past interaction behavior of others with domain expertise within that domain or the real-time engagement with domain experts (for example, in an instant messaging scenario). Exploratory search systems will leverage the search behaviors of other users within information spaces to provide recommendations about paths to follow and documents of interest. Recommendations can be made based upon vast amounts of historical log data generated by users of search engines. By employing data mining techniques, frequent patterns will emerge from logs that may be useful in support of exploration. For exploratory searches, there is novel value in encountering pages not frequently visited by other users of the system. These pages may contain information that yields unique insights or competitive advantage. However, a trade-off exists between the recommendation of documents that are difficult to find (and have small visit counts) and relevance to the current search task, since pages visited infrequently for this task may only be partially relevant. Nonetheless, page recommendation algorithms may foster serendiptitous information discoveries based upon the prior exploration of others with similar information needs. Existing search and browsing interfaces could allow users to mark useful landmarks as they encounter them, with a view to support later revisitation by them or by those with related interests.

6.1.8 New Evaluation Paradigms

The deployment of exploratory search interfaces at Web scale opens up new opportunities for their evaluation. In this lecture, we have discussed the evaluation of such interfaces by using small numbers of users engaged in laboratory settings or in longitudinal, naturalistic studies. The ability to monitor the use of exploratory search applications by thousands, if not millions, of users, allows system designers to monitor how these systems are used and adapt their components to suit the needs of their user population. Exploratory search may occur over multiple search sessions, and it can be difficult to evaluate exporatory search system effectiveness in a laboratory setting. The creation of shared data sets, pluggable search, indexing, and interaction components (such as crawling mechanisms, ranking algorithms, and interaction logging instrumentation) will cut the lag time from conception to implementation for system developers. More effective evaluation methodologies will improve the quality of the exploratory search systems that reach users.

6.2 CONCLUDING REMARKS

In this lecture, we have defined exploratory search, tied the concept to extant research in related communities, proposed a set of features of exploratory search systems and requirements for their evaluation, and presented a vision for the future of the field. The volume of related work demonstrates the breadth of interest in helping users develop enhanced mental capacities through search technology. The desirable features of exploratory search systems described in Chapter 4 exist presently in separate search systems. There is an opportunity to merge these features in an exploratory search application (or plug-in), capable of federating search requests to external search providers where appropriate. This application would augment existing search systems and assist exploratory searchers when requested or when typical exploratory search behavior, such as extensive topic-related browsing, is detected.

The future prospects for exploratory search are bright. Although the search community is still defining Exploratory Search 1.0, interest in the area continues to expand, and technological advances are making novel ideas (e.g., immersive environments, holographic projections, and mobile computing) a reality. Through interdisciplinary collaborations, academic and industrial researchers can make rapid progress toward the aspirations of the visionaries who started the research community on the journey toward human intelligence amplification. Intellectually oriented, problem-solving humans, engaged in activities such as scientific discovery, will be the first beneficiaries of these advances. However, search technology will evolve rapidly beyond its current form, and exploratory search applications will be as pervasive as Web search engines are today, if not more so. At that point, the dreams of Bush, Licklider, Engelbart, Nelson, and others of an "enlightened society" may actually be realized.

* * * *

Acknowledgments

The many participants in the exploratory search workshops and journal special issues have helped further research in this field. We would like to thank Gary Marchionini for composing this lecture series and for his vision in the exploratory search community. We also thank Daniel Tunkelang for his comments and suggestions on a draft of this lecture, and Diane Cerra for her assistance in getting this lecture to press.

References

Agichtein, E., Brill, E., and Dumais, S. T. (2006). Improving web search ranking by incorporating user behavior information. In *Proceedings of the 29th Annual International ACM SIGIR Conference on Research and Development in Information Retrieval*, pp. 19–26.

Ahlberg, C., Williamson, C., and Shneiderman, B. (1992). Dynamic queries for information exploration: An implementation and evaluation. In *Proceedings of the ACM SIGCHI Conference on Human Factors in Computing Systems*, pp. 619–626.

Ahlberg, C., and Shneiderman, B. (1994). Visual information seeking: Tight coupling of dynamic query filters with starfield displays. In *Proceedings of the ACM SIGCHI Conference on Human Factors in Computing Systems*, pp. 313–317.

Allan, J. (2003). HARD Track Overview in TREC 2003: High accuracy retrieval from documents. In *Proceedings of the Text Retrieval Conference*, pp. 24–37.

Alonso, O., Rose, D.E., and Stewart, B. (2008). Crowdsourcing for relevance evaluation. *SIGIR Forum*, 42(2), pp. 10–16.

Anderson, J. R. (1990). *The Adaptive Character of Thought*. Hillsdale, NJ: Lawrence Erlbaum Associates.

Anderson, J. R., Bothell, D., Byrne, M. D., Douglass, S., Lebiere, C., and Qin, Y. (2004). An integrated theory of the mind. *Psychological Review*, 111(4), pp. 1036–1060.

Anderson, L. W., and Krathwohl, D. R. (Eds.). (2001). *A Taxonomy for Learning, Teaching, and Assessing: A Revision of Bloom's Taxonomy of Educational Objectives*. New York, NY: Longman.

Armstrong, R., Freitag, D., Joachims, T., and Mitchell, T. (1995). WebWatcher: A learning apprentice for the world wide web. In *Proceedings of the AAAI Spring Symposium on Information Gathering from Heterogeneous, Distributed Environments*, pp. 6–12.

Ashby, W. R. (1956). *An Introduction to Cybernetics*. London, UK: Chapman & Hall.

Attfield, S., Blandford, A., and Dowell, J. (2003). Information seeking in the context of writing: a design psychology interpretation of the 'problematic situation'. *Journal of Documentation*, 59(4), pp. 430–453.

Baeza-Yates, R., and Pino, J. A. (1997). A first step to formally evaluate collaborative work. In *Proceedings of the ACM SIGGROUP Conference on Supporting Group Work*, pp. 56–60.

Balabanovic, M., and Shoham, Y. (1995). Learning information retrieval agents: Experiments with automated web browsing. In *Proceedings of the AAAI Spring Symposium on Information Gathering from Heterogeneous, Distributed Environments*, pp. 13–18.

Bates, M. J. (1984). The fallacy of the perfect 30-item online search. *Research Quarterly*, 24(1), pp. 43–50.

Bates, M. J. (1986a). An exploratory paradigm for online information retrieval. In *Proceedings of the 6th International Research Forum in Information Science*, pp. 91–99.

Bates, M. J. (1986b). Subject access in online catalogs: A design model. *Journal of the American Society for Information Science*, 37(6), pp. 357–376.

Bates, M. J. (1989). The design of browsing and berrypicking techniques for the online search interface. *Online Review*, 13(5), pp. 407–424.

Bates, M. J. (1990). Where should the person stop and the information search interface start? *Information Processing and Management*, 26(5), pp. 575–591.

Bates, M. J. (2004). What is browsing—really? A model drawing from behavioural science research. *Information Research*, 12(4), paper 330. [Available at http://InformationR.net/ir/12-4/paper330.html]

Bawden, D. (1986). Information systems and the stimulation of creativity. *Journal of Information Science*, 12(5), pp. 203–216.

Belkin, N. J. (1978). Information concepts for information science. *Journal of Documentation*, 34(1), pp. 55–85.

Belkin, N. J. (2000). Helping people find what they don't know. *Communications of the ACM*, 43(8), pp. 59–61.

Belkin, N. J., Cool, C., Kelly, D., Lin, S.-J., Park, S. Y., Perez-Carballo, J., and Sikora, C. (2001). Iterative exploration, design and evaluation of support for query reformulation in interactive information retrieval. *Information Processing and Management*, 37(3), pp. 403–434.

Belkin, N. J., Cool, C., Stein, A., and Theil, U. (1993). Cases, scripts and information seeking strategies: On the design of interactive information retrieval systems. *Expert Systems with Applications*, 29(3), pp. 325–344.

Belkin, N. J., Oddy, R. N., and Brooks, H. (1982a). ASK for information retrieval: Part 1. *Journal of Documentation*, 38(2), pp. 61–71.

Belkin, N.J., Oddy, R. N., and Brooks, H. (1982b). ASK for information retrieval: Part 2. *Journal of Documentation*, 38(3), pp. 145–164.

Belkin, N. J., and Vickery, A. (1985). *Interaction in Information System: A Review of Research From Document Retrieval to Knowledge-Based System*, 188–198. London: The British Library.

Bell, W. J. (1991). *Searching Behaviour: The Behavioural Ecology of Finding Resources*. London, UK: Chapman & Hall.

Berlyne, D. E. (1960). *Conflict, Arousal and Curiosity*. New York, NY: McGraw-Hill.

Bharat, K. (2000). SearchPad: Explicit capture of search context to support web search. In *Proceedings of 9th International World Wide Web Conference on Computer Networks*, pp. 493–501.

Bier, E. A., Stone, M. A., Pier, K., Buxton, W., and DeRose, T. D. (1993). Toolglass and magic lenses: The see through interface. In *Proceedings of the 20th Annual ACM SIGGRAPH Conference on Computer Graphics and Interactive Techniques*, pp. 73–80.

Bilenko, M., and White, R. W. (2008). Mining the search trails of surfing crowds: Identifying relevant websites from user activity. In *Proceedings of the 17th International World Wide Web Conference*, pp. 51–60.

Bloom, B. S. (1956). *Taxonomy of Educational Objectives, Handbook I: The Cognitive Domain*. New York, NY: David McKay Co. Inc.

Borlund, P. (2003). The IIR evaluation model: A framework for evaluation of interactive information retrieval systems. *Information Research*, 8(3), paper 152 [Available at: http://informationr.net/ir/8-3/paper152.html]

Borlund, P., and Ingwersen, P. (1998). Measures of relative relevance and ranked half-life: Performance indicators for interactive IR. In *Proceedings of the 21st Annual International ACM SIGIR Conference on Research and Development in Information Retrieval*, pp. 324–331.

Brajnik, G., Guida, G., and Tasso, C. (1987). User modeling in intelligent information retrieval. *Information Processing and Management*, 23(4), pp. 305–320.

Budd, J. M. (2004). Relevance: language, semantics, philosophy. *Library Trends*, 52(3), pp. 447–462.

Budzik, J., and Hammond, K. J. (2000). User interactions with everyday applications as context for just-in-time information access. In *Proceedings of the Annual Conference on Intelligent User Interfaces*, pp. 44–51.

Bush, V. (1945). As we may think. *Atlantic Monthly*, July, pp. 101–108.

Byström, K., and Järvelin, K. (1995). Task complexity affects information seeking and use. *Information Processing and Management*, 31(2), pp. 191–213.

Campbell, I. (1999). Interactive evaluation of the ostensive model, using a new test collection of images with multiple relevance assessments. *Journal of Information Retrieval*, 2(1), pp. 89–114.

Campbell, I. (2000). The ostensive model of developing information needs. Unpublished doctoral dissertation, University of Glasgow, Glasgow, UK.

Campbell, I., and Van Rijsbergen, C. J. (1996). Ostensive model of information needs. In *Proceedings of the 2nd International Conference on Conceptions of Library and Information Science: Integration in Perspective*, pp. 251–268.

Capra, R., and Marchionini, G. (2008). The relation browser tool for faceted exploratory search. In *Proceedings of the 8th ACM/IEEE-CS Joint Conference on Digital Libraries*, p. 420.

Card, S., Mackinlay, J. D., and Shneiderman, B. (1999). *Readings in Information Visualization: Using Vision to Think*. San Francisco, CA: Morgan Kaufmann.

Card, S. K., Pirolli, P., Van der Wege, M., Morrison, J. B., Reeder, R. W., Schraedley, P. K., and Boshart, J. (2001). Information scent as a driver of web behavior graphs. In *Proceedings of the ACM SIGCHI Conference on Human Factors in Computing Systems*, pp. 498–505.

Chalmers, M., and Chitson, P. (1992). Bead: explorations in information visualization. In *Proceedings of the 15th Annual International ACM SIGIR Conference on Research and Development in Information Retrieval*, pp. 330–337.

Charnov, E. L. (1976). Optimal foraging: The marginal value theorem. *Theoretical Population Biology*, 9, pp. 129–136.

Chi, E. H. Pirolli, P., Chen, K., and Pitkow, J. (2001). Using information scent to model user information needs and actions on the web. In *Proceedings of the ACM SIGCHI Conference on Human Factors in Computing Systems*, pp. 490–497.

Choo, C. W., Detlor, B., and Tunbull, D. (2000). Information seeking on the web: An integrated model of browsing and searching. *FirstMonday*, 5(2). [Available from http://firstmonday.org/issues/issue5_2/choo/index.html].

Cleverdon, C. W., Mills, J., and Keen, E. M. (1966). An inquiry in testing of information retrieval systems. (2 vols.). Aslib Cranfield Research Project, College of Aeronautics, Cranfield, UK.

Cool, C., Belkin, N. J., and Koenemann, J. (1996). On the potential utility of negative relevance feedback in interactive information retrieval. In *Proceedings of the 19th Annual International ACM SIGIR Conference on Research and Development in Information Retrieval*, p. 341.

Corbett, A. T., Koedinger, K. R., and Anderson, J. R. (1997). Intelligent tutoring systems. In Helander, M. G., Landauer, T. K., and Prabhu, P. V. (Eds.) *Handbook of Human–Computer Interaction* (pp. 849–874). Amsterdam, the Netherlands: Elsevier Science.

Croft, W. B., and Thompson, R. H. (1987). I^3R: A new approach to the design of document retrieval systems. *Journal of the American Society for Information Science*, 38(6), pp. 389–404.

Cronen-Townsend, S., Zhou, Y., and Croft, W. B. (2002). Predicting query performance. In *Proceedings of the 25th Annual International ACM SIGIR Conference on Research and Development in Information Retrieval*, pp. 299–306.

Cruz-Neira, C., Sandin, D. J., DeFanti, T. A., Kenyon, R. C., and Hart, J. C. (1992). The CAVE: Audio visual experience automatic virtual environment. *Communications of the ACM*, 35(6), pp. 64–72.

Cutrell, E., Robbins, D., Dumais, S., and Sarin, R. (2006). Fast, flexible filtering with phlat. In *Proceedings of the ACM SIGCHI Conference on Human Factors in Computing Systems*, pp. 261–270.

Cutting, D.R., Pedersen, J.O., Karger, D., and Tukey, J.W. (1992). Scatter/Gather: A cluster-based approach to browsing large document collections. In *Proceedings of the 15th Annual*

International ACM SIGIR Conference on Research and Development in Information Retrieval, pp. 318–329.

Czerwinski, M., and Horvitz, E. (2002). An investigation of memory for daily computing events. In *Proceedings of the HCI Conference*, pp. 230–245.

Dervin, B. (1977). Useful theory for librarianship: Communication, not information. *Drexel Library Quarterly*, 13(3), pp. 16–32.

Dervin, B. (1992). From the mind's eye of the user: The sense-making qualitative–quantitative methodology. In Glazier, J. D., and Powell, R. R. (Eds.), *Qualitative Research in Information Management*, pp. 64–81.

Dervin, B. (1998). Sense-making theory and practice: An overview of user interests in knowledge seeking and use. *Journal of Knowledge Management*, 2(2), pp. 36–46.

Dervin, B., and Nilan, M. (1986). Information needs and uses. In: Williams, M.E. (Ed.) *Annual Review of Information Systems and Technology*, 21, pp. 3–33.

Dontcheva, M., Drucker, S., Wade, G., Salesin, D., and Cohen, M. (2006). Summarizing personal web browsing sessions. In *Proceedings of the 19th Annual ACM UIST Symposium on User Interface Software and Technology*, pp. 115–124.

Dumais, S., and Belkin, N. J. (2005). The TREC Interactive Track: Putting the user into search. In Voorhees, E., and Harman, D. (Eds.), *TREC: Experiment and Evaluation in Information Retrieval* (pp. 123–152). Cambridge, MA: MIT Press.

Dumais, S. T., Cutrell, E., Sarin, R., and Horvitz, E. (2004). Implicit queries (IQ) for contextualized search. In *Proceedings of the 27th Annual International ACM SIGIR Conference on Research and Development in Information Retrieval*, p. 594.

Dumais, S. T., Cutrell, E. Cadiz, J. J., Jancke, G., Sarin, R., and Robbins, D. C. (2003). Stuff I've seen: A system for personal information retrieval and re-use. In *Proceedings of the 26th Annual International ACM SIGIR Conference on Research and Development in Information Retrieval*, pp. 72–79.

Efthimiadis, E. N. (1996). Query expansion. In: Williams, M. E. (Ed.), *Annual Review of Information Systems and Technology*, 31, pp. 121–187.

Egan, D. E., Remde, J. R., Gomez, L. M., Landauer, T. K., Eberhardt, J., and Lochbaum, C. C. (1989). Formative design evaluation of superbook. *ACM Transactions on Information Systems*, 7(1), pp. 30–57.

Eliot, T. S. (1942). *Little Gidding*. London, UK: Faber and Faber.

Ellis, D. (1984). Theory and explanation in information retrieval research. *Journal of Information Science*, 8(1), pp. 25–38.

Ellis, D. (1989). A behavioural approach to information retrieval design. *Journal of Documentation*, 45(3), pp. 171–212.

Ellis, D., Cox, D., and Hall, K. (1993). A comparison of the information seeking patterns of researchers in the physical and social sciences. *Journal of Documentation*, 49(4), pp. 356–369.

Ellis, D., and Haugan, M. (1997). Modeling the information seeking patterns of engineers and research scientists in an industrial environment. *Journal of Documentation*, 53(4), pp. 384–403.

Engelbart, D. (1962). *Augmenting Human Intellect: A Conceptual Framework*. Summary Report AFOSR-3233. Menlo Park, CA: Stanford Research Institute.

Finkelstein, L., Gabrilovich, E., Matias, Y., Rivlin, E., Solan, Z., Wolfman, G., and Ruppin, E. (2002). Placing search in context: The concept revisited. *ACM Transactions on Information Systems*, 20(1), pp. 116–131.

Foltz, P., and Landauer, T. (2007). Helping people find and learn from documents: Exploiting synergies between human and computer retrieval with SuperManual. In Landauer, T. K., McNamara, D. S., Dennis, S., and Kintsch, W. (Eds.), *The Handbook of Latent Semantic Analysis*, pp. 323–345. Mahwah, NJ: Erlbaum.

Ford, N. (1980). Relating information needs to learner characteristics in higher education. *Journal of Documentation*, 36, pp. 165–191.

Ford, N. (1999). Information retrieval and creativity: Towards support for the original thinker. *Journal of Documentation*, 55(5), pp. 528–542.

Foster, A., and Ford, N. (2003). Serendipity and information seeking: An empirical study. *Journal of Documentation*, 59(3), pp. 321–340.

Furnas, G. W. (1986). Generalized fisheye views. In *Proceedings of the ACM SIGCHI Conference on Human Factors in Computing Systems*, pp. 16–23.

Furnas, G. W. (1997). Effective view navigation. In *Proceedings of the ACM SIGCHI Conference on Human Factors in Computing Systems*, pp. 367–374.

Garfield, E. (1970). When is a negative search result positive? *Essays of an Information Scientist*, 1, pp. 117–118.

Glover, E. J., Lawrence, S., Birmingham, W. P., and Giles, C. (1999). Architecture of a metasearch engine that supports user information needs. In *Proceedings of the 8th ACM CIKM Conference on Information and Knowledge Management*, pp. 210–216.

Golovchinsky, G. (1997). What the query told the link: the integration of hypertext and information retrieval. In *Proceedings of the 8th ACM Conference on Hypertext*, pp. 67–74.

Golovchinsky, G., Qvarfordt, P., and Pickens, J. (2009). Collaborative information seeking. *IEEE Computer*, 42(3), in press.

Grimes, C., Tang, D., and Russell, D. (2007). Query logs alone are not enough. In *Proceedings of WWW 2007 Workshop on Query Log Analysis: Social and Technological Changes*.

Harman, D. (1993). Overview of the first TREC conference. In *Proceedings of the 16th Annual ACM SIGIR Conference of Research and Development in Information Retrieval*, pp. 36–47.

Harter, S. P. (1992). Psychological relevance for information science. *Journal of the American Society for Information Science*, 43(9), pp. 602–615.

Hearst, M. (1995). TileBars: Visualization of term distribution information in full text information access. In *Proceedings of the ACM SIGCHI Conference on Human Factors in Computing Systems*, pp. 59–66.

Hearst, M. A. (2006). Clustering versus faceted categories for information exploration. *Communications of the ACM*, 49(4), pp. 59–61.

Hearst, M. A., and Pedersen, J. O. (1996). Reexamining the cluster hypothesis: Scatter/gather on retrieval results. In *Proceedings of 19th Annual International ACM SIGIR Conference on Research and Development in Information Retrieval*, pp. 76–84.

Heer, J., and Chi, E. H. (2001). Identification of web user traffic composition using multi-modal clustering and information scent. In *Proceedings of the Workshop on Web Mining, SIAM Conference on Data Mining*, pp. 51–58.

Henry, W. M., Leigh, J. A., Tedd, L. A., and Williams, P. W. (1980). *Online Searching: An Introduction*. London, UK: Butterworth.

Hersh, W. R., Turpin, A., Price, S., Chan, B., Kraemer, D., Sacherek, L., and Olson, D. (2000). Do batch and user evaluation give the same results? In *Proceedings of the 23rd Annual International ACM SIGIR Conference on Research and Development in Information Retrieval*, pp. 17–24.

Hinckley, K., Zhao, S., Sarin, R., Baudisch, P., Cutrell, E., Shilman, M., and Tan, D. (2007). InkSeine: In situ search for active note taking. In *Proceedings of the ACM SIGCHI Conference on Human Factors in Computing Systems*, pp. 251–260.

Horvitz, E., Breese, J., Heckerman, D., Hovel, D., and Koos R. (1998). The Lumiere project: Bayesian user modeling for inferring the goals and needs of software users. In *Proceedings of the 14th Conference on Uncertainty in Artificial Intelligence*, pp. 256–265.

Hughes, R. N. (1997). Intrinsic exploration in animals: motives and measurement. *Behavioural Processes*, 41(3), pp. 213–226.

Ingwersen, P. (1992). *Information Retrieval Interaction*. London, UK: Taylor Graham.

Ingwersen, P. (1994). Polyrepresentation of information needs and semantic entities: Elements of a cognitive theory for information retrieval interaction. In *Proceedings of the 17th Annual International ACM SIGIR Conference on Research and Development in Information Retrieval*, pp. 101–111.

Ingwersen, P. (1996). Cognitive perspectives of information retrieval interaction: Elements of a cognitive IR theory. *Journal of Documentation*, 52(1), pp. 3–50.

Ingwersen, P. (2001). Cognitive information retrieval. In: Williams, M. (Ed.), *Annual Review of Information Science and Technology*, 34, pp. 3–51.

Ingwersen, P. (2002). Cognitive perspectives of document representation. In *Proceedings of the 4th International Conference on Conceptions of Library and Information Science*, pp. 285–300.

Ingwersen, P., and Järvelin, K. (2005). *The Turn: Integration of Information Seeking and Retrieval in Context*. Secaucus, NJ: Springer-Verlag.

Ingwersen, P., and Pejtersen, A. M. (1986). User requirements: Empirical research and information systems design. In: Ingwersen, P., Kajberg, L., and Pejtersen, A. M. (Eds.), *Information Technology and Information Use: Towards a Unified View of Information and Information Technology* (pp. 111–124). London, UK: Taylor Graham.

Ingwersen, P., and Wormell, I. (1987). Improved subject access, browsing and scanning mechanisms in modern online IR. In *Proceedings of the 9th Annual International ACM SIGIR Conference on Research and Development in Information Retrieval*, pp. 68–75.

Ingwersen, P., and Wormell, I. (1989). Modern indexing and retrieval techniques matching different types of information needs. In: Koskiala, S., and Launo, R. (Eds.) *Information, Knowledge, Evolution* (pp. 79–90). London, UK: North-Holland.

Joachims, T., Granka, L., Pan, B., Hembrooke, H., and Gay, G. (2005). Accurately interpreting clickthrough data as implicit feedback. In *Proceedings of the 28th Annual International ACM SIGIR Conference on Research and Development in Information Retrieval*, pp. 154–161.

Jones, R., Rey, B., Madani, O., and Greiner, W. (2006). Generating query substitutions. In *Proceedings of the 15th Annual International World Wide Web Conference*, pp. 387–396.

Jones, W. (2008). *Keeping Found Things Found: The Study and Practice of Personal Information Management*. Burlington, MA: Morgan Kaufmann.

Käki, M. (2005). Findex: Search result categories help users when document rankings fail. In *Proceedings of ACM SIGCHI Conference on Human Factors in Computing Systems*, pp. 131–140.

Keller, R., Wolf, S., Chen, J., Rabinowitz, J., and Mathe, N. A. (1997). Bookmarking service for organizing and sharing URLs. *Computer Networks and ISDN Systems*, 29(8–13), pp. 1103–1114.

Kelly, D., and Belkin, N. J. (2004). Display time as implicit feedback: understanding task effects. In *Proceedings of the 27th Annual International ACM SIGIR Conference on Research and Development in Information Retrieval*, pp. 377–384.

Kelly, D., Dumais, S., and Pedersen, J. (2009). Evaluation challenges and directions for information seeking support systems. *IEEE Computer*, 42(3), in press.

Kelly, D., and Teevan, J. (2003). Implicit feedback for inferring user preference: A bibliography. *ACM SIGIR Forum*, 37(2), pp. 18–28.

Kim, S., and Soergel, D. (2005). Selecting and measuring taskcharacteristics as independent variables. In *Proceedings of the 68th Annual Meeting of the American Society for Information Science and Technology*, 42, Information Today, Medford, NJ.

Kleiboemer, A. J., Lazear, M. B., and Pedersen, J. O. (1996). Tailoring a retrieval system for naïve users. *In Proceedings of the 5th Annual Symposium on Document Analysis and Information Retrieval.*

Klein, G., Moon, B., and Hoffman, R. F. (2006). Making sense of sensemaking I: Alternative perspectives. *IEEE Intelligent Systems*, 21(4), pp. 70–73.

Kleinberg, J. (1999). Authoritative sources in a hyperlinked environment. *Journal of the ACM*, 46(5), pp. 604–632.

Koenemann, J., and Belkin, N. J. (1996). A case for interaction: A study of interactive information retrieval behavior and effectiveness. In *Proceedings of ACM SIGCHI Conference on Human Factors in Computing Systems*, pp. 205–212.

Kuhlthau, C. C. (1991). Inside the search process: Information seeking from the user's perspective. *Journal of the American Society for Information Science*, 42(5), pp. 361–371.

Kuhlthau, C. C. (1993). *Seeking Meaning*. Norwood, NJ: Ablex.

Kuhlthau, C. C. (2004). *Seeking Meaning: A Process Approach to Library and Information Services (2nd Edition)*. Westport, CT: Libraries Unlimited.

Kuhn, T. S. (1970). *The Structure of Scientific Revolutions (2nd Edition)*. Chicago, IL: University of Chicago Press.

Kules, B., and Capra, R. (2008). Creating exploratory tasks for a faceted search interface. In *Proceedings of 2nd Workshop on Human–Computer Interaction*, pp. 18–21.

Kules, B., and Shneiderman, B. (2008). Users can change their web search tactics: Design guidelines for categorized overviews. *Information Processing and Management*, 44(2), pp. 463–484.

Kwasnik, B. H. (1992). A descriptive study of the functional components of browsing. In *Proceedings of the IFIP TC2/WG2.7 Working Conference on Engineering for Human–Computer Interaction*, pp. 191–203.

Kwasnik, B. H. (1999). The role of classification in knowledge representation and discovery, *Library Trends*, 48(1), pp. 22–47.

Lagergren, E., and Over, P. (2001). Comparing interactive information retrieval systems across sites: The TREC-6 interactive track matrix experiment. In *Proceedings of the 21st Annual International ACM SIGIR Conference on Research and Development in Information Retrieval*, pp. 164–172.

Landauer, T. K. (2002). On the computational basis of learning and cognition: Arguments from LSA. *The Psychology of Learning and Motivation*, 41, pp. 43–84.

Licklider, J. C. R. (1960). Man–computer symbiosis. *IRE Transactions on Human Factors in Electronics*, v. HFE-1, pp. 4–11.

Lieberman, H. (1995). Letizia: An agent that assists web browsing. In *Proceedings of the 14th International Joint Conference on Artificial Intelligence*, pp. 475–480.

Lieberman, H., Fry, C., and Weitzman, L. (2001). Exploring the web with reconnaissance agents. *Communications of the ACM*, 44 (7), pp. 69–75.

Lin, J., and Smucker, M. D. (2008). How do users find things with PubMed?: Towards automatic utility evaluation with user simulations. In *Proceedings of 31st Annual International ACM SIGIR Conference on Research and Development in Information Retrieval*, pp. 19–26.

Loeb, S., and Terry, D. B. (1992). Information filtering: preface to the special section. *Communications of the ACM*, 35(12), pp. 26–28.

Loewenstein, G. (1994). The psychology of curiosity: A review and reinterpretation. *Psychological Bulletin*, 116(1), pp. 75–98.

Lynch, K. (1960). *The Image of the City*. Cambridge, MA: MIT Press.

Mackay, D. M. (1960). What makes the question? *The Listener*, 62, pp. 789–790.

Maglio, P. P., Barrett, R., Campbell, C. S., and Selker, T. (2000). SUITOR: An attentive information system. In *Proceedings of the 5th Annual IUI Conference on Intelligent User Interfaces*, pp. 169–176.

Mann, T. (1987). *A Guide to Library Research Methods*. New York, NY: Oxford University Press.

Marchionini, G. (1995). *Information Seeking in Electronic Environments*. New York, NY: Cambridge University Press.

Marchionini, G. (2006a). Exploratory search: From finding to understanding. *Communications of the ACM*, 49(4), pp. 41–46.

Marchionini, G. (2006b). Toward human–computer information retrieval. *Bulletin of the American Society for Information Science*, 32(5), pp. 20–22.

Marchionini, G., and Brunk, B. (2003). Toward a general relation browser: A GUI for information architects. *Journal of Digital Information*, 4(1); jodi.ecs.soton.ac.uk/Articles/v04/i01/Marchionini/.

Marchionini, G., and Shneiderman, B. (1988). Finding facts vs. browsing knowledge in hypertext systems. *IEEE Computer*, 21(1), pp. 70–79.

Marchionini, G., and White, R. W. (2009). Information-seeking support systems: Introduction to theme issue. *IEEE Computer*, 42(3), in press.

Marshall, C. C., and Shipman, F. M. (1995). Spatial hypertext: Designing for change. *Communications of the ACM*, 38(8), pp. 88–97.

Maslow, A. H. (1954). *Motivation and Personality*. New York, NY: Harper and Row.

Mellon, C. A. (1986). Library anxiety: A grounded theory and its development. *College and Research Libraries*, 47, pp. 160–165.

Millen, D., Feinberg, J., and Kerr, B. (2005). Social bookmarking in the enterprise. *Queue*, 3(9), pp. 28–35.

Miller, G. A. (1983). Informavores. In Machlup, F., and Mansfield, U. (Eds.) *The Study of Information: Interdisciplinary Messages* (pp. 111–113). New York, NY: Wiley-Interscience.

Mitchell, T., Caruana, R., Freitag, D., McDermott, J., and Zabowski, D. (1994). Experience with a learning personal assistant. *Communications of the ACM*, 37 (7), pp. 81–91.

Morris, D., Morris, M. R., and Venolia, G. (2008). SearchBar: A search-centric web history for task resumption and information re-finding. In *Proceedings of ACM SIGCHI Conference on Human Factors in Computing Systems*, pp. 1207–1216.

Morris, M. R., and Horvitz, E. (2007). SearchTogether: An interface for collaborative web search. In *Proceedings of the 20th Annual ACM UIST Symposium on User Interface Software and Technology*, pp. 3–12.

Nelson, T. H. (1965). Complex information processing: a file structure for the complex, the changing and the indeterminate. In *Proceedings of the 20th National ACM Conference*, pp. 84–100.

Newell, A., and Simon, H. A. (1972). *Human Problem Solving*. Englewood Cliffs, NJ: Prentice-Hall.

Noerr, P. L., and Noerr, K. T. B. (1985). Browse and navigate: An advance in database access methods. *Information Processing and Management*, 21(3), pp. 205–213.

O'Brien, H. (2008). Defining and Measuring Engagement in User Experiences with Technology. Unpublished doctoral dissertation, Dalhousie University, Halifax, Canada.

O'Day, V., and Jeffries, R. (1993). Orienteering in an information landscape: how information seekers get from here to there. In *Proceedings of the ACM SIGCHI Conference on Human Factors in Computing Systems*, pp. 438–445.

Oddy, R. N. (1977). Information retrieval through man–machine-dialogue. *Journal of Documentation*, 33(1), pp. 1–14.

Pace, S. (2004). A grounded theory of the flow experiences of web users. *International Journal of Human–Computer Studies*, 60(3), pp. 327–363.

Pao, M. (1993). Term and citation searching: A field study. *Information Processing and Management*, 29(1), 95–112.

Patterson, E. S., Roth, E. M., and Woods, D. D. (2001). Predicting vulnerabilities in computer-supported inferential analysis under data overload. *Cognition Technology and Work*, 3(4), pp. 224–237.

Piaget, J. (1978). *The Development of Thought: Equilibration of Cognitive Structures*. New York, NY: Viking Penguin.

Pickens, J., Golovchinsky, G., Shah, C., Qvarfordt, P., and Back, M. (2008). In *Proceedings of the 31st ACM SIGIR Conference on Research and Development in Information Retrieval*, pp. 315–322.

Pirolli, P. (1997). Computational models of information scent-following in a very large browsable text collection. In *Proceedings of the ACM SIGCHI Conference on Human Factors in Computing Systems*, pp. 3–10.

Pirolli, P. (2007). *Exploratory Search Systems*. Retrieved from http://web.mac.com/peter.pirolli/Professional/Blog/Entries/2007/5/18_Exploratory_Search_Systems.html on December 15, 2008.

Pirolli, P. (2008). *Information Foraging Theory: Adaptive Interaction with Information*. New York, NY: Oxford University Press.

Pirolli, P. (2009). Powers of 10: Modeling complex information seeking systems at multiple scales. *IEEE Computer*, 42(3), in press.

Pirolli, P., and Card, S. (1995). Information foraging in information access environments. In *Proceedings of the ACM SIGCHI Conference on Human Factors in Computing Systems*, pp. 51–58.

Pirolli, P., and Card, S. (1999). Information foraging. *Psychology Review*, 106(4), pp. 643–675.

Pirolli, P., and Card, S. K. (2005). The sensemaking process and leverage points for analyst technology as identified through cognitive task analysis. In *Proceedings of the International Conference on Intelligence Analysis*.

Pirolli, P., and Rao, R. (1996). Table lens as a tool for making sense of data. In *Proceedings of the Working Conference on Advanced Visual Interfaces*, pp. 67–80.

Pirolli, P., Schank, P., Hearst, M., and Diehl, C. (1996). Scatter/gather browsing communicates the topic structure of a very large text collection. In *Proceedings of the ACM SIGCHI Conference on Human Factors in Computing Systems*, pp. 213–220.

Plaisant, C. (2004). The challenge of information visualization evaluation. In *Proceedings of the Working Conference on Advanced Visual Interfaces*, pp. 109–116.

Pollitt, A. S., Ellis, G. P., and Smith, M. P. (1994). HIBROWSE for bibliographic databases. *Journal of Information Science*, 20(6), pp. 413–426.

Qu, Y., and Furnas, G. (2007). Model-driven formative evaluation of exploratory search: A study under a sensemaking framework. *Information Processing and Management*, 44(2), pp. 534–555.

Rao, R., and Card, S. K. (1994). The table lens: Merging graphical and symbolic representations in an interactive focus + context visualization for tabular information. In *Proceedings of the ACM SIGCHI Conference on Human Factors in Computing Systems*, pp. 318–322.

Rich, E. (1983). Users are individuals: Individualizing user models. *International Journal of Human–Computer Studies*, 51, pp. 323–338.

Ringel, M., Cutrell, E., Dumais, S., and Horvitz, E. (2003). Milestones in time: the value of landmarks in retrieving information from personal stores. In *Proceedings of the IFIP TC13 International Conference on Human–Computer Interaction*, pp. 184–191.

Russell, D. M., Stefik, M. J., Pirolli, P. L., and Card, S. K. (1993). The cost structure of sensemaking. In *Proceedings of ACM SIGCHI Conference on Human Factors in Computing Systems*, pp. 269–276.

Ruthven, I. (2008). Interactive information retrieval. *Annual Review of Information Science and Technology*, 42, pp. 43–91.

Salton, G., and Buckley, C. (1990). Improving retrieval performance by relevance feedback. *Journal of the American Society for Information Science*, 41(4), pp. 288–297.

Salton, G., and McGill, J. M. (1983). *Introduction to Modern Information Retrieval.* New York, NY: McGraw-Hill.

Saracevic, T. (1996). Relevance reconsidered '96. In *Proceedings of the 2nd International Conference on Conceptions of Library and Information Sciences*, pp. 201–218.

Saracevic, T. (1997). The stratified model of information retrieval interaction: Extension and applications. In *Proceedings of the American Society for Information Science Annual Meeting*, 34, pp. 313–327.

Saracevic, T. (2007). Relevance: A review of the literature and a framework for thinking on the notion in information science. Part II: Nature and manifestations of relevance. *Journal of the American Society for Information Science and Technology*, 58(13), pp. 1915–1933.

Saracevic, T., Kantor, P., Chamis, A., and Trivison, D. (1988). A study of information seeking and retrieving. I. Background and methodology. *Journal of the American Society for Information Science*, 39(3), pp. 161–176.

schraefel, m. c. (2009). Building knowledge: What's beyond keyword search? *IEEE Computer*, 42(3), in press.

schraefel, m. c., Karam, M., and Zhao, S. (2003). Listen to the music: Audio preview cues for exploration of online music. In *Proceedings of the IFIP TC13 International Conference on Human–Computer Interaction*, pp. 192–195.

schraefel, m. c., Smith, D. A., Owens, A., Russell, A., Harris, C., and Wilson, M. L. (2005). The evolving mSpace platform: Leveraging the semantic web on the trail of the Memex. In *Proceedings of the 16th ACM Conference on Hypertext and Hypermedia*, pp. 174–183.

schraefel, m. c., Wilson, M., Russell, D., and Smith, D. A. (2006). mSpace: improving information access to multimedia domains with multimodal exploratory search. *Communications of the ACM*, 49(4), pp. 47–49.

schraefel, m. c., Zhu, Y., Modjeska, D., Wigdor, D., and Zhao, S. (2002). Hunter gatherer: interaction support for the creation and management of within-web-page collections. In *Proceedings of the 11th Annual World Wide Web Conference*, pp. 172–181.

Shen, X., Tan, B., and Zhai, C. (2005). Context-sensitive information retrieval using implicit feedback. In *Proceedings of the 28th ACM SIGIR Conference on Research and Development in Information Retrieval*, pp. 43–50.

Shneiderman, B., Byrd, D., and Croft, W. B. (1997). Clarifying search: A user-interface framework for text searches. *D-Lib Magazine.*

Simon, H. A. (1971). Designing organizations in an information-rich world. In Greenberger, M. (Ed.), *Computers, communications, and the public interest* (pp. 37–53). Baltimore, MD: Johns Hopkins University Press.

Simon, H. A. (1973). The structure of ill-structured problems. *Artificial Intelligence*, 4(3), pp. 181–204.

Smeaton, A. F., Lee, H., Foley, C., McGivney, S., and Gurrin, C. (2006). Fischlár-diamondtouch: Collaborative video searching on a table. In *Multimedia Content Analysis, Management, and Retrieval*, pp. 64–74.

Smith, S., Glenberg, A., and Bjork, R. (1978). Environmental context and human memory. *Memory and Cognition*, 6(4), pp. 342–353.

Smyth, B., Balfe, E., Boydell, O., Bradley, K., Briggs, P., Coyle, M., and Freyne, J. (2005). A live-user evaluation of collaborative web search. In *Proceedings of the International Joint Conference on Artificial Intelligence*, pp. 1419–1424.

Soergel, D. (1999). The rise of ontologies or the reinvention of classification, *Journal of the American Society for Information Science and Technology*, 50(12), pp. 1119–1120.

Spink, A., Griesdorf, H., and Bateman, J. (1998). From highly relevant to not relevant: Examining different regions of relevance. *Information Processing and Management*, 34(5), pp. 599–621.

Stefik, M. J., Baldonado, M. Q. W., Bobrow, D., Card, S., Everett, J., Lavendel, G., Marimon, D., Newman, P., Russell, D., and Smoliar, S. (1999). *The knowledge sharing challenge: The sense-making white paper*. PARC, Inc.

Stephens, D. W., and Krebs, J. R. (1986). *Foraging Theory*. Princeton, NJ: Princeton University Press.

Stoica, E., and Hearst, M. (2004). Nearly automated metadata hierarchy creation. In *Proceedings of the Annual Conference of the North American Chapter of the Association for Computational Linguistics (Companion Volume)*, pp. 117–120.

Tague-Sutcliffe, J. (1992). Measuring the informativeness of a retrieval process. In *Proceedings of the 15th Annual International ACM SIGIR Conference on Research and Development in Information Retrieval*, pp. 23–36.

Tang, R., and Solomon, P. (1998). Towards an understanding of the dynamics of relevance judgements: An analysis of one person's search behaviour. *Information Processing and Management*, 43(2/3), pp. 237–256.

Taylor, A. G. (1992). *Introduction to Cataloging and Classification (8th Edition)*. Englewood, CO: Libraries Unlimited.

Taylor, R. S. (1968). Question negotiation and information seeking inlibraries. *College and Research Libraries*, 29(3), pp. 178–194.

Teevan, J., Alvarado, C., Ackerman, M. S., and Karger, D. R. (2004). The perfect search engine is not enough: A study of orienteering behavior in directed search. In *Proceedings of the ACM SIGCHI Conference on Human Factors in Computing Systems*, pp. 415–422.

Teevan, J., Adar, E., Jones, R., and Potts, M. (2007). Information re-retrieval: Repeat queries in Yahoo's logs. In *Proceedings of the 30th Annual International ACM SIGIR Conference on Research and Development in Information Retrieval*, pp. 151–158.

Thorndyke, P. W., and Goldin, S. E. (1983). *Spatial learning and reasoning skill.* In Pick, H. L., and Acredolo, L. P. (Eds.) *Spatial Orientation: Theory, Research, and Application* (pp. 195–217). New York, NY: Plenum Press.

Tombros, A., and Sanderson, M. (1998). Advantages of query biased summaries in information retrieval. In *Proceedings of the 21st Annual International ACM SIGIR Conference on Research and Development in Information Retrieval*, pp. 2–10.

Toms, E. G., O'Brien, H., Kopak, R., and Freund, L. (2005). Searching for relevance in the relevance of search. In *Proceedings of the 5th International Conference on Conceptions of Library and Information Sciences*, pp. 59–78.

Trigg, R. H. (1988). Guided tours and tabletops: tools for communicating in a hypertext environment. In *Proceedings of the ACM CSCW Conference on Computer-Supported Cooperative Work*, pp. 216–226.

Tukey, J. W. (1977). *Exploratory Data Analysis.* Reading, MA: Addison-Wesley.

Tulving, E. (1983). *Elements of Episodic Memory.* New York, NY: Oxford University Press.

Tunkelang, D. (2009). Precision AND Recall. *IEEE Computer*, 42(3), in press.

Vakkari, P. (1999). Task complexity, problem structure and information actions: Integrating studies on information seeking and retrieval. *Information Processing and Management*, 35(6), pp. 819–837.

Vakkari, P., and Hakala, N. (2000). Changes in relevance criteria and problem stages in task performance. *Journal of Documentation*, 56(5), pp. 540–562.

Van Rijsbergen, C. J. (1979). *Information Retrieval.* Butterworth: London, UK.

Vickery, A., and Brooks, H. M. (1987). PLEXUS: The expert system for referral. *Information Processing and Management*, 23(2), pp. 99–117.

Viégas, F. B., Wattenberg, M., van Ham, F., Kriss, J., and McKeon, M. (2007). Many eyes: A site for visualization at internet scale. *IEEE Transactions on Visualization and Computer Graphics*, 13(6), pp. 1121–1128.

Voorhees, E. M. (2007). Building test collections for adaptive information and retrieval: What to abstract for what cost? In *Proceedings of the 1st International Workshop on Adaptive Information Retrieval*, p. 12.

Voorhees, E. M., and Harman, D. K. (2005). *TREC: Experiment and Evaluation in Information Retrieval.* Cambridge, MA: MIT Press.

Vosniadou, S., and Brewer, W. F. (1987). Theories of knowledge restructuring in development. *Review of Educational Research*, 57(1), pp. 51–67.

Wade, S. G., and Willett, P. (1988). INSTRUCT: A teaching package for experimental methods in information retrieval. Part III—Browsing, clustering and query expansion, *Program*, 22(1), pp. 44–61.

Wang, T., Plaisant, C., Quinn, A. J., Stanchak, R., Murphy, S., and Shneiderman, B. (2008). Aligning temporal data by sentinel events: discovering patterns in electronic health records. In *Proceedings of the ACM SIGCHI Conference on Human Factors in Computing Systems*, pp. 457–466.

Wexelblat, A., and Maes, P. (1999). Footprints: History-rich tools for information foraging. In *Proceedings of the ACM SIGCHI Conference on Human Factors in Computing Systems*, pp. 270–277.

White, R. W., Bilenko, M., and Cucerzan, S. (2007). Studying the use of popular destinations to enhance web search interaction. In *Proceedings of 30th Annual International ACM SIGIR Conference on Research and Development in Information Retrieval*, pp. 159–166.

White, R. W., and Drucker, S. M. (2007). Investigating behavioral variability in web search. In *Proceedings of the 16th Annual World Wide Web Conference*, pp. 21–30.

White, R. W., Dumais, S. T., and Teevan, J. (2009). Characterizing the influence of domain expertise on web search behavior. In *Proceedings of the 2nd Annual International ACM WSDM Conference on Web Search and Data Mining*, in press.

White, R. W., Kules, B., and Bederson, B. (2005). Exploratory search interfaces: Categorization, clustering and beyond: report on the XSI 2005 workshop at the Human–Computer Interaction Laboratory, University Of Maryland. *SIGIR Forum*, 39(2), pp. 52–56.

White, R. W., Kules, B., Drucker, S., and schraefel., m. c. (2006a). Supporting exploratory search: Introduction to special section. *Communications of the ACM*, 49(4), pp. 36–39.

White, R. W., and Marchionini, G. (2007). Examining the effectiveness of real-time query expansion. *Information Processing and Management*, 43(3), pp. 685–704.

White, R. W., and Morris, D. (2007). Investigating the querying and browsing behavior of advanced search engine users. In *Proceedings of the 30th Annual International ACM SIGIR Conference on Research and Development in Information Retrieval*, pp. 255–262.

White, R. W., Muresan, G., and Marchionini, G. (2006b). Report on ACM SIGIR 2006 workshop on evaluating exploratory search systems. *SIGIR Forum*, 40(2), pp. 52–60.

White, R. W., Ruthven, I., and Jose, J. M. (2005b). A study of factors affecting the utility of implicit relevance feedback. In *Proceedings of the 28th Annual International ACM SIGIR Conference on Research and Development in Information Retrieval*, pp. 35–42.

White, R. W., Ruthven, I., Jose, J. M., and Van Rijsbergen, C. J. (2005a). Evaluating implicit feedback models using searcher simulations. *ACM Transactions on Information Systems*, 23(3), pp. 325–361.

White, R. W., Song, H., and Liu, J. (2006c). Concept maps to support oral history search and use. In *Proceedings of the 6th ACM/IEEE-CS Joint Conference on Digital Libraries*, pp. 192–193.

Wigdor, D., Forlines, C., Baudisch, P., Barnwell, J., and Shen, C. (2007). LucidTouch: A see-through mobile device. In *Proceedings of the 20th Annual ACM UIST Symposium on User Interface Software and Technology*, pp. 269–278.

Wilson, A. (2005). PlayAnywhere: A compact interactive tabletop projection-vision system. In *Proceedings of the 18th Annual ACM UIST Symposium on User Interface Software and Technology*, pp. 83–92.

Wilson, T. D. (1997). Information behaviour: An interdisciplinary perspective. *Information Processing and Management*, 33(4), pp. 551–572.

Wilson, T. D. (1999). Models in information behavior research. *Journal of Documentation*, 55(3), pp. 249–270.

Wolfe, J. M. (1994). Guided search 2.0: A revised model of visual search. *Psychonomic Bulletin and Review*, 1(2), pp. 202–238.

Xu, J., and Croft, W. B. (2000). Improving the effectiveness of information retrieval with local context analysis. *ACM Transactions on Information Systems*, 18(1), pp. 79–112.

Yee, K. P., Swearingen, K., Li, K., and Hearst, M. (2003). Faceted metadata for image search and browsing. In *Proceedings of the ACM SIGCHI Conference on Human Factors in Computing Systems*, pp. 401–408.

Zamir, O., and Etzioni, O. (1999). Grouper: A dynamic clustering interface to Web search results. In *Proceedings of the 8th Annual International World Wide Web Conference*, pp. 1361–1374.

Zhang, P., Soergel, D., Klavans, J. L., and Oard, D. W. (2008). Extending sense-making model with ideas from cognition and learning theories. In *Proceedings of the American Society for Information Science and Technology Conference*, in press.

Zipf, G. K. (1949). *Human Behavior and the Principle of Least Effort*. Cambridge, MA: Addison-Wesley.

Zloof, M. (1975). Query-by-example: the invocation and definition of tables and forms. In *Proceedings of the 1st International Conference on Very Large Data Bases*, pp. 1–24.

Author Biographies

Ryen William White has been a researcher in the Text Mining, Search, and Navigation Group at Microsoft Research, Redmond since May 2006. He was a faculty research associate in the Institute for Advanced Computer Studies at the University of Maryland, College Park from 2004 to 2006, working on research projects funded by the National Science Foundation (NSF) and United States Department of Defense.

He received his Ph.D. in Computer Science from the University of Glasgow, United Kingdom in October 2004. His doctoral dissertation entitled "Implicit Feedback for Interactive Information Retrieval" focused on the development and testing of novel methods to help people search electronic environments such as the World Wide Web. In recognition of his dissertation research, he received the British Computer Society's "Distinguished Dissertation" award for the best Computer Science Ph.D. dissertation in the United Kingdom in 2004–2005.

Ryen's research interests include exploratory search, implicit relevance feedback, query expansion, log analysis, and the evaluation of search systems with humans and simulations. He has coauthored over 80 conference and journal publications in these areas. He was a recipient of best paper award at the 2003 *INTERACT Conference*, the best student paper award at the 2004 *European Conference on Information Retrieval*, and the best paper award at the 2007 *ACM SIGIR Conference on Research and Development in Information Retrieval.*

He has cochaired numerous workshops on exploratory search in conjunction with ACM SIGIR and ACM SIGCHI conferences, as well as workshops sponsored by the NSF, and some that were independently funded. He guest coedited the April 2006 special section of *Communications of the ACM* on supporting exploratory search, a special issue of *Information Processing and Management* on evaluating exploratory search systems in March 2008, and a theme issue of *IEEE Computer* on information-seeking support systems in March 2009.

Ryen served as posters chair of the 2008 European Conference on Information Retrieval and as treasurer of the 2008 ACM Conference on Information and Knowledge Management. He has served on proposal evaluation panels for the European Research Council and the NSF, and serves on the editorial board of the Journal of Information Retrieval.

Resa Ariel Roth is an Honors College graduate of Washington State University, United States, where she earned her B.S. degree in 2006. She is an experienced technical writer with a desire to expand the human mind through the use of progressive search technology.

Made in the USA